BARRON'S

Painless Statistics

Patrick Honner,
M.A., M.S.Ed.

For Rachele, the center of my data.

Published by Kaplan, Inc., d/b/a Barron's Educational Series
1515 W Cypress Creek Road
Fort Lauderdale, FL 33309
www.barronseduc.com

ISBN: 978-1-5062-8158-2

10 9 8 7 6 5 4 3 2 1

Kaplan, Inc., d/b/a Barron's Educational Series print books are available at special quantity discounts to use for sales promotions, employee premiums, or educational purposes. For more information or to purchase books, please call the Simon & Schuster special sales department at 866-506-1949.

Contents

How to Use This Book

Painless statistics? It might sound impossible, but it's not. Statistics is easy . . . or at least it can be with the help of this book!

Whether you are learning statistics for the first time, or you are revisiting ideas you may have forgotten, this book is for you. It provides a clear introduction to statistics that is both fun and instructive. Don't be afraid. Dive in—it's painless!

Painless Icons and Features

This book is designed with several unique features to help make learning statistics easy.

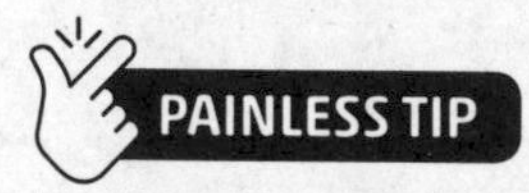

You will see Painless Tips throughout the book. These include helpful tips, hints, and strategies on the surrounding topics.

Caution boxes will help you avoid common pitfalls or mistakes. Be sure to read them carefully.

1+2=3 MATH TALK

These boxes translate "math talk" into plain English to make it even easier to understand statistics.

Reminders will call out information that is important to remember. Each reminder will relate to the current chapter or will reference key information you learned in a previous chapter.

BRAIN TICKLERS

There are brain ticklers throughout each chapter in the book. These quizzes are designed to make sure you understand what you've just learned and to test your progress as you move forward in the chapter. Complete all the Brain Ticklers and check your answers. If you get any wrong, make sure to go back and review the topics associated with the questions you missed.

PAINLESS STEPS

Complex procedures are divided into a series of painless steps. These steps help you solve problems in a systematic way. Follow the steps carefully, and you'll be able to solve most statistics problems.

EXAMPLES

Most topics include examples. If you are having trouble, research shows that writing or copying the problem may help you understand it.

ILLUSTRATIONS

Painless Statistics is full of illustrations to help you better understand statistics topics. You'll find graphs, charts, and other instructive illustrations to help you along the way.

SIDEBARS

These shaded boxes contain extra information that relates to the surrounding topics. Sidebars can include detailed examples or practice tips to help keep statistics interesting and painless.

Chapter Breakdown

Chapter One gives you a painless introduction to statistics, with an example that shows you where data comes from, what you can do with it, and what it can do for you.

Chapter Two is all about data, the object of study in statistics. You'll learn about the different types of data and how to represent and visualize it with charts and graphs.

Chapter Three shows you the power of statistics. You'll see how questions about large sets of data can be answered by considering just a handful of representative numbers using descriptive statistics like the mean and standard deviation.

Chapter Four introduces you to the important distributions you'll encounter when working with data and shows you how you can use properties of those distributions to make analyzing that data as painless as possible.

Chapter Five focuses on a single distribution of data: the normal distribution. Normally distributed data pops up everywhere in the world—from test scores to biological measurements and so much more—so knowing how to work with the normal distribution is key to applying your statistical knowledge.

Chapter Six introduces you to probability, the mathematics of likelihood. A firm grasp of basic probability helps you put your data in context and is essential to understanding important ideas in statistics, like sampling and inference.

Chapter Seven extends your study of probability into conditional probability. You will learn how to analyze the impact that events can have on each other, another central theme in applying statistical knowledge.

Chapter Eight develops the basics of statistical sampling and sampling distributions, which allow you to use statistics to understand entire populations by just looking at a small sample of the data.

Chapter Nine shows you how to construct and interpret confidence intervals, which are important statistical tools that allow you to make educated guesses about unknown quantities while accounting for the uncertainty inherent in the process.

Chapter Ten teaches you about statistical significance: what it is, what it isn't, and how you can measure it.

Chapter Eleven gives you the basic statistical tools you need to tackle bivariate data, like scatterplots, correlation, and linear regression.

Chapter Twelve puts your statistical knowledge in context by helping you interpret and understand the statistical claims you'll see and hear all around you.

Chapter 1

An Introduction to Data

Statistics is the study of data. Data is everywhere, and statistics can help you collect it, categorize it, analyze it, and interpret it. Data comes in many forms—it can be a series of stock prices, the number of touchdowns, or a list of favorite ice creams—but for all these forms, you can use statistics to help make sense of the data. It can be hard to understand a large set of numbers, but with statistics, working with data can be painless.

An Example of Working with Data

How can data help you make decisions and better understand your world? Here's a simple example that illustrates what statistics can do for you and just how painless it can be.

Example:

Imagine that you live in one city and work in another. To get from home to work, you have two different commuting options. Commute A involves riding three different trains and making two transfers. Commute B involves riding just one train, but it also involves a slightly longer walk from the station to your job.

Unsure of which commute is better, you try both for a while and record how long it takes you to get from home to work. In other words, you collect some data.

Commute	Length of Commute (in minutes)
A	32, 43, 26, 27, 42, 30, 42, 44, 31, 30, 27, 46
B	34, 36, 40, 41, 37, 35, 37, 41, 39, 37, 36, 41, 37, 40, 39

These are your two data sets. A *data set* is a set of information, and in this example, the information comes in the form of numbers. Each individual number in these data sets is a *data point*, and each data point represents the length of a commute in minutes. There are 12 data points in the first data set and 15 data points in the second set. The variable n is used to indicate the size of a data set. In the first set, $n = 12$, and in the second set, $n = 15$.

So, what can you do with this data? Naturally, you might be interested in knowing which commute is better, but how can you compare the 12 numbers in the first data set with the 15 numbers in the second data set? This is where statistics come in.

A *statistic* is a number that captures some feature or characteristic of a data set. A good statistic communicates an important or useful piece of information about the data, and the beauty of a statistic is that it's just a single number. Instead of trying to think about the entire data set all at once, you can focus only on the statistic.

PAINLESS TIP

A statistic is a way of thinking about an entire data set by considering just one number. This makes it easier to understand and analyze your data set.

There are many different statistics you can use to help make sense of data. The first one you should know is a familiar statistic that describes where the "center" of the data is.

The Center of a Data Set

One of the first steps toward understanding a data set is determining where its center is. A common way to measure the center of a data set is by finding the mean of the data. The *mean* of a set of numbers is the sum of all those numbers divided by the size of the data set. The mean is often referred to as the *average* because, in a sense, the mean can be thought of as an average representative of the data set. For example, people around the world typically range in height from 1 foot tall to 7 feet tall, but the mean height is around 5.5 feet. You might think of an average person being around 5.5 feet tall.

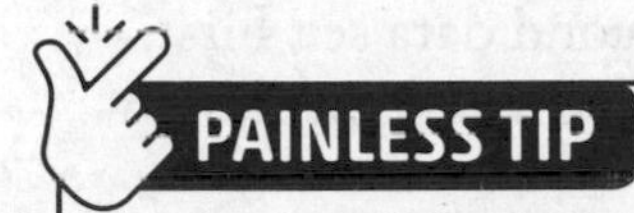

The word *average* has different meanings in different contexts. Sometimes *average* refers to the mean; other times it refers to other statistics (such as the median), as discussed further in Chapter 3.

To compute the mean of a data set, simply follow these painless steps.

Step 1: Add all the values in the data set.

Step 2: Divide that sum by the number of values in the data set.

The mean you get by adding up all the numbers in a set and dividing by the number of values is also known as the *arithmetic mean* of a set of numbers.

For example, to find the mean of the set of commute times for Commute A, first add up all the numbers in that data set:

$$32 + 43 + 26 + 27 + 42 + 30 + 42 + 44 + 31 + 30 + 27 + 46 = 420$$

Then divide the sum by 12, the number of values in the set:

$$\frac{420}{12} = 35$$

The result is 35, so the mean of this data set is 35 minutes. The unit is minutes because each data point in the set was measured in minutes. So in the calculation of the mean, you divided a sum of minutes, 420, by a pure number, 12.

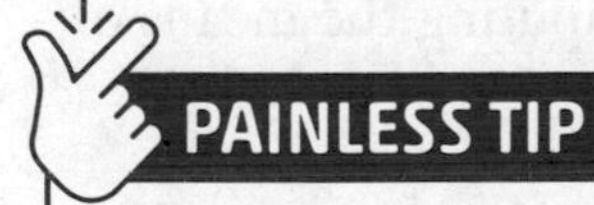

Don't be afraid to use your calculator! You are expected to use technology to help with calculations in statistics, especially when the data sets are large. Many calculators have statistics capabilities that allow you to enter in a list of numbers and compute the mean with a single command.

Here's the computation of the mean for the second data set. First, add up all the numbers in the data set, and then divide that sum by 15 (the number of values in the set):

$$\frac{34+36+40+41+37+35+37+41+39+37+36+41+37+40+39}{15}$$

$$=\frac{570}{15}=38$$

Therefore, the mean commuting time for Commute B is 38 minutes.

CAUTION—Major Mistake Territory!

The mean of a data set is *not* necessarily a data point in the set. For example, the mean of the second data set is 38 minutes, but 38 is not a number in that data set. The mean can be thought of as an average member of the data set, but it's not required to be a member of the set.

If your ultimate goal is to compare the two commutes, the means that you just calculated allow you to do exactly that. Each data set is now represented by a single number, so instead of trying to compare all the numbers in the first set with all the numbers in the second set, you can just compare the means. This tells you that, on average, Commute A takes 35 minutes and Commute B takes 38 minutes, so Commute A is, on average, shorter than Commute B.

The mean is an example of a summary statistic. A *summary statistic* is just what it sounds like: it summarizes the data in a single number. This frees you from the burden of having to think about all the data at once. You'll encounter more summary statistics (that describe other features of data) in the next section and in Chapter 3, but first try the following Brain Ticklers to practice computing the means of data sets.

REMINDER

Comparing two numbers is much easier than comparing two large data sets. This is one of the benefits of using statistics.

BRAIN TICKLERS Set #1

1. Compute the mean of each of the following data sets.

   a. 34, 18, 21, 65, 9, 11, 18, 52

   b. 17, 27, −14, 3, 0, −25, −28, −3, 32

   c. 5, 5, 5, 5, 5, 5, 5, 5

2. Create a data set with $n = 5$ that has mean 10.
3. Suppose a data set with $n = 4$ has mean 100. A new data point is added to the data set, and the new mean is 101. What is the new data point?
4. Suppose a data set has mean 100, and you know that one data point is 75. Explain why there must be another data point in the data set that is greater than 100.

(Answers are on page 14.)

The Spread of a Data Set

Now that you've compared the means of the two sets of commute times, you might be tempted to conclude that Commute A is better because, on average, it is shorter than Commute B. This is a reasonable conclusion to draw, but there's more to this story, and more summary statistics will help tell it.

As a statistic, the mean captures some information about the data but not all of it. Other summary statistics capture other features that might also be important to understanding the data. For instance, take another look at the two data sets of commute times. This time, the data points are arranged in order from smallest to largest.

Commute A: 26, 27, 27, 30, 30, 31, 32, 42, 42, 43, 44, 46

Commute B: 34, 35, 36, 36, 37, 37, 37, 37, 39, 39, 40, 40, 41, 41, 41

Ordering the data like this makes it easy to spot the smallest and largest numbers in each data set. These are also useful statistics. For example, Commute A has a minimum time of 26 minutes and a maximum time of 46 minutes, while Commute B has a minimum time of 34 minutes and a maximum time of 41 minutes. This says something useful about the data.

Range

The *range* of a data set is the difference between the largest number in the set and the smallest number in the set. For example, the range of the data from Commute A is 46 − 26 = 20 minutes, while the range of the data from Commute B is 41 − 34 = 7 minutes. Range offers a second opportunity to compare the data sets. Comparing the ranges shows that the data from Commute A is more spread out than the data from Commute B.

Range is another summary statistic, and it captures something different about the data than the mean does. The mean measures the *central tendency* of a data set, that is, what the center of the data set looks like. The range measures the *dispersion*, or spread, of a set of data. These two different kinds of summary statistics form a powerful combination. Knowing where the center is as well as how far the data extends gives you a lot of information about the nature of the data.

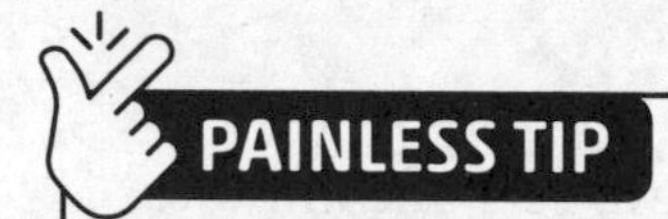

A single statistic offers a single piece of information about a data set, so using different statistics together creates a fuller picture of the data. Depending on the nature of the data, just a few important pieces of information can be enough to tell you what you want to know.

Pairing these statistics with a visual representation of the data can make the information even clearer. The illustrations that follow are dot plots of the two data sets of commute times. A *dot plot* is a visual representation of the frequency of the data: for each data point in the data set, a dot is placed over its corresponding number. For example, there are two 30s in the data for Commute A, so there are two dots above the 30 in the dot plot on the left. Likewise, since there are four 37s in the data for Commute B, there are four dots above 37 in the dot plot on the right.

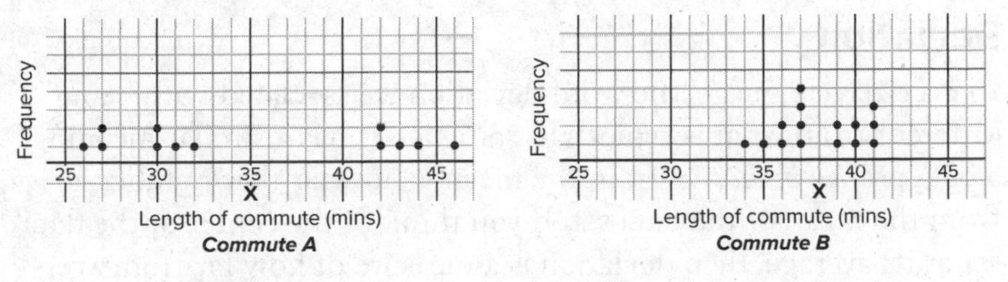

Figure 1–1 **Figure 1–2**

Visualizing data in plots and graphs is a great way to understand the data; you'll learn about different ways to picture data throughout this book. Visualizations can also help put the summary statistics in context. For example, the mean is located near the center of each dot plot (an X has been added to indicate the location of the mean), so you can see where the "center" of the data is. Also, you can see that the dot plot for Commute A is much wider than the dot plot for Commute B, which is what the range tells you about the data.

So, how does the range help you answer the question "Which commute is better?" Comparing the means showed that Commute A is, on average, shorter than Commute B, but the larger range of the data for Commute A shows that there is more *variation* in those commute times than in the times for Commute B. With Commute A, you may have a commute of 26 minutes, which is much shorter than average, but you also might have a commute of 46 minutes, which is much longer than average. On average, you will probably get to work faster with Commute A than with Commute B, but there's a chance you might get there very early or very late with Commute A.

With Commute B, your average commute time is slightly longer, but there's less variation in your commute times. According to the data, the commute is never longer than 41 minutes or shorter than 34 minutes. These commute times are more steady and predictable.

Range gives you an idea of the variation in a data set (i.e., how much the data varies), but there are summary statistics that capture this variation more precisely. These are statistics that measure deviation.

Deviation

In everyday language, the word deviation is associated with being different from what is typical. In statistics, it has a similar meaning: a data point's *deviation* from the mean is how far that number is from the mean of the data set. If you think of the center of the data set as its average, then deviation is a measure of how far from average a data point is.

For example, the first data point for Commute A is 32 minutes. Since the mean commute time is 35 minutes, the 32-minute commute differs from the mean commute by 3 minutes. This deviation can be computed by subtracting the data value from the mean: $35 - 32 = 3$ minutes. You can also subtract the mean from the data value, but in this case, you get a negative number because the value is less than the mean: $32 - 35 = -3$ minutes. In statistics, you usually want to think of deviations as distances, so you want them to be positive numbers. Taking the *absolute value* of the deviation will guarantee that your deviations are never negative.

> The absolute value of a number is its distance to 0. For example, the distance from 7 to 0 is 7, so the absolute value of 7, denoted $|7|$, is 7. Similarly, the distance from –3 to 0 is 3, so the absolute value of –3, or $|-3|$, is 3. In short, the absolute value of a number is what you get when you ignore its sign.

In the case of the data point 32, the absolute value of the deviation is 3, so it doesn't matter if you compute it as $|32 - 35| = |-3| = 3$ or $|35 - 32| = |3| = 3$.

This is an example of computing the deviation of an individual data point, but you can also measure the deviation of an entire data set. The trick is to compute all the individual deviations and then find their mean. This statistic is called the *mean absolute deviation* (MAD), which measures the average distance to the mean in a data set. It is also simply called the *mean deviation*.

To compute the mean absolute deviation for a data set, follow these painless steps.

Step 1: Compute the mean of the data set.

Step 2: Calculate the absolute value of each individual data point's deviation from the mean. This will form a new data set.

Step 3: Compute the mean of this new data set of deviations.

For example, to compute the mean absolute deviation of the data set from Commute A, recall that the mean is 35, so you've already completed the first step. Next, you need to calculate the absolute value of each individual data point's deviation from the mean.

32, 43, 26, 27, 42, 30, 42, 44, 31, 30, 27, 46

Commute A Data

The deviation for the first data point is $35 - 32 = 3$; for the second point, it's $35 - 43 = -8$ (which you take the absolute value of to get $|-8| = 8$); for the third point, it's $35 - 26 = 9$; and so on. This yields a new set of numbers—the deviations from the mean of the original set:

3, 8, 9, 8, 7, 5, 7, 9, 4, 5, 8, 11

Finally, find the mean of this new data set. To do so, add up all the individual deviations, and then divide by the number of deviations:

$$\frac{3+8+9+8+7+5+7+9+4+5+8+11}{12} = 7$$

Thus, the mean absolute deviation of the data from Commute A is 7 minutes. As with the mean of this data set, the unit of mean absolute deviation is also minutes, since all the individual deviations are measured in minutes and the sum is divided by a pure number.

The mean deviation says that, on average, a data point is 7 minutes away from the mean. So in Commute A, the average commute time is 35 minutes, but also on average, a commute will take 7 minutes, more or less, than 35 minutes.

This can sound confusing at first: is the average commute time 35 minutes, or $35 - 7$ minutes, or $35 + 7$ minutes? Remember, however, that the two statistics measure different features of the data. The mean says that the center of the data is at 35, but not every commute will be 35 minutes long. The commute times are spread out around the center, and the mean deviation measures that spread.

Just as means provide a useful way to compare and contrast two data sets, so do mean deviations. Here's the data for Commute B, this time ordered from least to greatest.

$$34, 35, 36, 36, 37, 37, 37, 37, 39, 39, 40, 40, 41, 41, 41$$

Commute B Data

The mean of this data set, as calculated earlier in this chapter, is 38. So the next step in computing the mean deviation is calculating the absolute value of the difference between each data point and 38:

$$4, 3, 2, 2, 1, 1, 1, 1, 1, 1, 2, 2, 3, 3, 3$$

Finally, compute the average of these individual deviations, which is

$$\frac{4+3+2+2+1+1+1+1+1+1+2+2+3+3+3}{15} = 2$$

So, the mean absolute deviation of the data set for Commute B is 2 minutes, which is considerably less than the mean deviation for Commute A. What does that say about the data? With Commute A, since the average deviation from the mean is 7 minutes, it would not be unusual for a commute to take around $35 + 7 = 42$ minutes. With Commute B, however, such a long commute would be much less likely, even though the average commute time is longer. This is because with Commute B, the average deviation from the mean of 38 minutes is only 2 minutes.

Notice that this all fits with the visualization of the data from the dot plots. For Commute A, the spread of the data around the mean is much wider, while for Commute B, the data is clumped together closer to its mean. This spread around the mean is what deviation is meant to represent, and mean absolute deviation is a simple statistic for capturing this characteristic of the data. In Chapter 3, you'll learn about a more common way to measure the deviation of a data set, the *standard deviation*. Calculating standard deviation is a little more complicated than calculating mean deviation, but the idea is exactly the same: it's just a way to measure how far, on average, data is from the center.

BRAIN TICKLERS Set #2

1. Compute the mean absolute deviation of each of the following data sets.

   a. 10, 20, 30, 40, 50, 60, 70, 80, 90, 100

   b. 17, 27, −14, 3, 0, −25, −28, −3, 32

   c. 5, 5, 5, 5, 5, 5, 5, 5

2. Construct a data set with $n = 10$, a mean of 20, and a MAD of 0.
3. Construct a data set with $n = 10$, a mean of 20, and a MAD of 3.
4. Suppose the MAD of a data set is 5 and that one of the data points has an individual deviation of 1. Explain how you know that some other data point must have an individual deviation greater than 5.

(Answers are on page 14.)

Interpreting Data

After recording a few weeks' worth of commute data and computing some summary statistics, what can you do with all this information? One obvious answer is that you can use it to decide which commute you should use to get to work in the future. Assuming future commutes are similar to past commutes, you can use the data you've collected and analyzed to draw inferences about what future commutes will be like and then make decisions based on those inferences. This is an example of a very common application of statistics, that is, trying to understand a large set of data (a data set of all possible commute times) by studying a small sample of the data (like your commute times from the past few weeks). You'll learn more about this kind of sampling and inference making later in this book.

If the goal is to make an inference, then which commute is better? Ultimately, that's up to you to decide, but summary statistics like mean and deviation have helped clarify the answer for you. Would you rather have a commute that is, on average, shorter but less predictable (like Commute A), or would you rather have a commute that is, on average, longer but more predictable (like Commute B)? As is often the case in statistics, there isn't necessarily a right or wrong answer. If you don't like surprises, you might choose the more predictable Commute B. If you're willing to risk occasionally being late for a shorter-on-average commute, you might choose Commute A. What works best for you depends on your needs and preferences.

Another reason statistics rarely provide clear right or wrong answers is that there's always more to the story than what your data tells you. When you apply statistics to study data, you may find that there are often other factors that haven't been accounted for. What hasn't been considered in analyzing these two commutes? Well, different commuting methods may have different costs. A person might have good reason to avoid multiple train transfers or long walks outside. Maybe there's a big penalty for being late or a big reward for being early. As useful as statistics like the mean and deviation are in comparing these two commutes, there's a lot of information they don't capture about the situation. It's important to remember that when you are interpreting data and drawing conclusions.

The data you collect and the statistics you choose to compute will always tell only part of the story. Choosing to work with some data (like commute times) and ignoring other data (like commute prices) impacts the kinds of analysis you can do and the kinds of conclusions you can safely draw. Replacing a large data set for a few statistics is a good trade in that it makes analyzing data more convenient, but it invariably leaves out some information. It's important to always be mindful of those choices, and those trade-offs, when working with data.

REMINDER

Every choice you make in statistics, from what data you collect to what techniques you apply, leaves some information behind. Math is often seen as a right-or-wrong discipline, but when you apply statistics to analyze data, conclusions are rarely clear-cut.

The commute times example illustrated how a few simple statistics can help you make sense of data. Knowing about the center and spread of a data set can help you make an informed decision about your morning commute, your investment opportunities, and much more. Throughout this book, you'll learn many more techniques for organizing, representing, analyzing, and interpreting data and for applying statistics the painless way.

BRAIN TICKLERS Set #3

1. Suppose you want to invest your money in an automobile company, and you decide to buy stock in whichever company has the highest stock price. Why might stock price not be the best data to focus on?
2. High schools are often ranked and compared based on the number of students who take Advanced Placement courses. What information might not be captured when focusing on this particular data?
3. If you wanted to use statistics to better understand your family's eating habits, what kinds of data might you collect?

(Answers are on page 15.)

Brain Ticklers—The Answers

Set #1, page 5

1. a. 28.5

   b. 1

   c. 5

2. There are many possible sets you could create. Examples include: {8, 9, 10, 11, 12}; {0, 0, 0, 0, 50}; {10, 10, 10, 10, 10}.
3. 105
4. If all the numbers in the data set were less than 100, then the mean of the data set would have to be less than 100. So, if one number in the data set is less than the mean, some number in the data set must be greater than the mean for the data set to "average out" to 100.

Set #2, page 11

1. a. 25

   b. $16\frac{2}{3}$ or 16.67

   c. 0

2. {20, 20, 20, 20, 20, 20, 20, 20, 20, 20}
3. There are several possibilities. Examples include: {17, 17, 17, 17, 17, 23, 23, 23, 23, 23} and {15, 16, 17, 18, 19, 21, 22, 23, 24, 25}.
4. MAD is the average deviation from the mean, so MAD is itself an average. If all the individual deviations were less than 5, the mean of those deviations would be less than 5. Therefore, if the mean of those deviations is 5 and one of the individual deviations is less than 5, then at least one individual deviation must be greater than 5.

Set #3, page 13

1. Stock price may not be the best indicator of the overall worth of a company. Different companies issue different amounts of stock, so stock price alone doesn't indicate the total value of a company. Plus, when it comes to investing, it might be more important to know things like how much revenue a company generates, how much debt it owes and if the stock pays dividends, among many other factors.
2. This particular data ignores other advanced, but non-AP, courses students might be taking. It also ignores the different kinds of opportunities that are available to students, such as in the arts, sports, or technical trades.
3. For this type of analysis, there are dozens of characteristics you could collect data about. Examples include how much money is spent on groceries per week, how much money is spent at restaurants, the average amount of calories per meal, the number of calories from vegetables, the number of meals and snacks per day, and so on.

Chapter 2

Data and Representations

Statistics is all about working with data, so you need to understand what data is. In this chapter, you'll learn about different types of data and how to work with them. You'll also learn about different ways of visualizing and representing data, which can help make analyzing and interpreting data painless.

Types of Data

Data is information. The information can be about people, places, or processes, and it can come in the form of numbers or characteristics. All these different kinds of information are different types of data, and there are two important types of data you should be familiar with.

Quantitative Data vs. Categorical Data

The data sets that were discussed in Chapter 1 were commute times, so the data came in the form of numbers. That is an example of quantitative data. *Quantitative data* is numeric information, like times, heights, prices, and test scores. This is a very common type of data in the application of statistics, and most of the examples you'll see in this book involve quantitative data.

However, data isn't always quantitative. Suppose that, instead of recording how long their daily commute is, you asked people the method they use to commute to work each day. In this scenario, the information you're looking for isn't a number; it's a mode of transportation (by car, by bus, by bike, and so on). This is an example of categorical data. *Categorical data* indicates the category, or categories, individuals belong to.

Even though categorical data isn't necessarily quantitative in nature, you can still use numbers to help understand it. For example, you might represent this categorical data in a table.

Commute Method	Percentage
Car	32%
Bus/Train	40%
Bike	16%
Walk	12%

A table is a way to represent the information and is helpful in understanding and interpreting the data. The table above, for example, shows what percentage of individuals fall into each category. You'll learn other ways to represent categorical data later in this chapter.

Whether the data is quantitative or categorical, it's always important to understand who the information is about. Data often describes people, but it can also describe objects or things (such as products, countries, or organizations). These objects or things are referred to as the *individuals* described by the data. When you encounter data, you should always ask the question, Who is being described by the data? Here's a simple example to get you thinking about the kinds of questions you should ask yourself when analyzing categorical data.

Example 1:

Refer to the following table, which represents some categorical data.

Favorite Movie Type	Percentage
Action/Adventure	24%
Comedy	38%
Horror	14%
Romance	16%
Documentary	8%

This data tells you about the kinds of movies people like, but you should ask yourself, Who were the individuals being asked about their favorite movie type? What this data means, and how it can be used, depends on the answer to this question. If this information came from American high school students, you might draw one conclusion from the data, but if it came from Canadian grandparents, you might draw a different conclusion.

Knowing who the data represents is part of understanding how the data is defined. Each characteristic captured in data is a *variable*, and it is very important to understand exactly what defines a variable. In the example from Chapter 1, does commute time start when you walk out your front door, or does it start when you step on the train? When it comes to commute type, does working from home count as walking to work? When discussing movie preferences, is *Thor: Ragnarok* considered an action movie or a comedy? Understanding the details of your data is essential to making proper sense of it.

When you start working with a data set, make sure you understand which individuals are described by the data and how the variables that make up the data are defined.

Populations and Samples

Sometimes when working with data, you collect all the information that exists about a subject and use statistics to analyze and study it. Usually, though, you can't actually collect all the data about a particular subject because there's just too much of it. In these situations, you instead study samples of data.

Imagine you are interested in understanding the movie preferences of all American adults. It isn't possible to ask every single American adult about movie preferences, so instead you might ask a small group of people what their preferences are. The goal is to use statistics to understand the smaller data set in hopes of understanding the larger one. In this situation, the smaller group is called a *sample*, and the larger data set that the sample comes from is the *population*. The data points you look at are *observations*. Similar to how a zoologist might observe a small group of wild animals to understand how the entire population behaves, you can think of each data point or sample as an observation of what an individual's movie preferences are.

CAUTION—Major Mistake Territory!

A population and a sample are *not* one and the same! If you are working with the entire set of data, that's the population. If you are working with a smaller set of data that represents the population, that's a sample. It's always important to distinguish between a population and a sample when applying statistical techniques.

Just as it's important to understand who the data describes and how the data is defined, it's equally important to understand where your samples came from. You'll learn much more about using samples to understand populations in Chapter 8.

Numerical information that represents a sample is called a *statistic*. Numerical information that represents a population is called a *parameter*.

BRAIN TICKLERS Set #4

1. Is the following data quantitative or categorical?

   a. The grade point averages of high school students

   b. The political affiliation of adult American women

   c. The available vacation days of federal employees

2. Consider a large corporation as a set of individuals to be studied. First, provide an example of a quantitative variable that relates to this set of individuals. Then, provide an example of a categorical variable for this set of individuals.

3. Suppose you were collecting data on the average amount of money held in savings accounts by Americans. What differences might you expect to see in the data if the individuals were American college students versus American adults between the ages of 45 and 55?

(Answers are on page 41.)

Representations of Data

Representing and visualizing data effectively is an important step toward understanding it. Here are some of most common and useful ways to represent data.

Pie Charts and Bar Charts

You can't browse the Internet without encountering pie charts and bar charts. These graphs are simple ways of representing categorical data, and they help you compare categories to each other and to the whole.

In a *pie chart*, each "slice of pie" represents a category. Here's an example that shows you what a pie chart representing categorical data looks like.

Example 2:

Recall the following data about commute methods.

Commute Method	Percentage
Car	32%
Bus/Train	40%
Bike	16%
Walk	12%

Now, here's that same data presented in a pie chart.

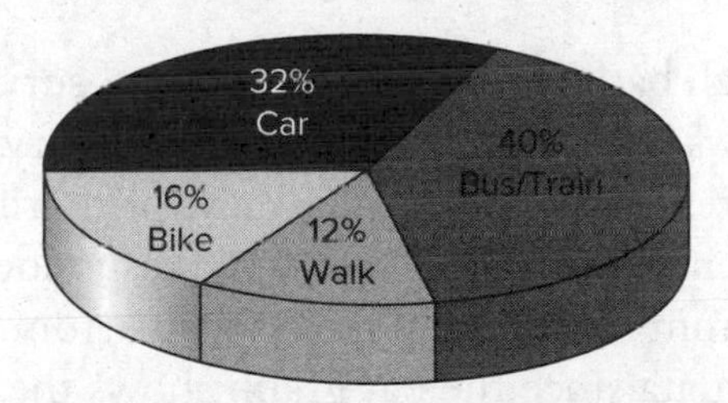

Figure 2–1. Method of Commute to Work

This pie chart helps you see that the amount of people who take the bus or train is comparable to the amount of people who drive a car. It also lets you easily see that those two categories cover almost three-quarters of all of the data. The vast majority of people represented here use one of those options to commute to work.

A bar chart is another way to visualize categorical data. A *bar chart* is a graph where the categories are listed across the horizontal axis and the height of each bar indicates the frequency of each category. The frequency could be given as a count or as a percentage.

Example 3:

Suppose a survey is conducted to determine how adult Americans get their news. Here's a bar chart that shows the results of that survey.

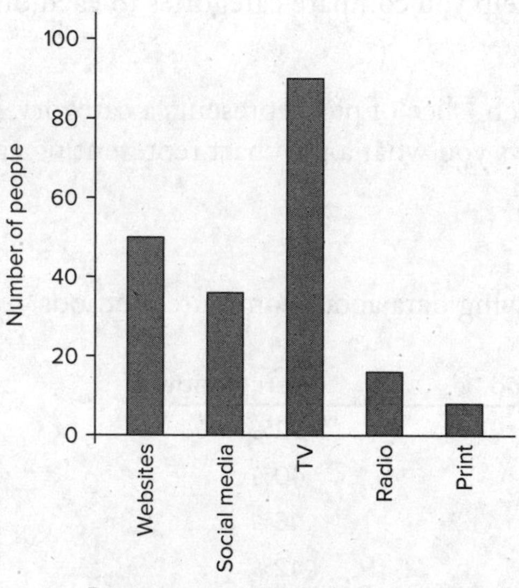

Figure 2–2

The height of each bar indicates the number of survey respondents who get their news primarily through each category. This bar chart tells you that 50 respondents get their news primarily from websites, 36 respondents from social media, 90 respondents from TV, 16 respondents from radio, and 8 respondents from print. This is an example of *count* data since the bar graph shows the count of data points in each category.

The title of this bar chart includes $n = 200$, which indicates that a total of 200 American adults were surveyed. You can also determine this by adding together the counts, which are just the heights of the bars. In this example, $50 + 36 + 90 + 16 + 8 = 200$. This works because each data point in the data set is in exactly one category.

Instead of showing raw counts, a bar chart can also show percentages. To find the percentage of data in each category, just divide the raw count in each category by the total number of values in the data set, which in this case is 200. Doing this changes each value from a frequency to a *relative frequency*. The height of each bar now shows the frequency of each value relative to the entire data set.

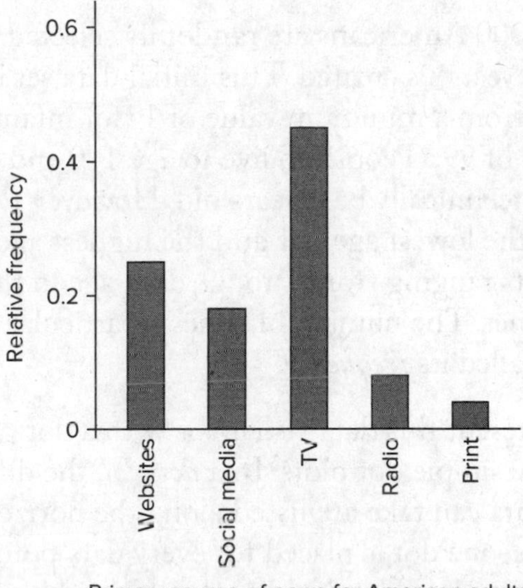

Figure 2–3

Notice that the overall shape of the bar chart is still the same as the one prior. Since you are dividing each count by the same amount, each bar is getting scaled by the same factor, so their relative sizes are unchanged. In a bar chart that indicates frequency, like this one, the heights of the bars will add to 1, or 100%.

CAUTION—Major Mistake Territory!

A lot of errors are made because graphs and charts are mislabeled or misread. Make sure your graph and axes are labeled appropriately when making a visualization of data, and make sure you read carefully when interpreting one.

Dot Plots and Histograms

Bar charts and pie charts are useful in visualizing categorical data. For visualizing quantitative data, dot plots and histograms are the most commonly used graphs. Here's an example that will walk you through these two important statistical tools.

Example 4:

Imagine that 1,000 Americans are randomly selected, and a data set of their ages (in years) is created. This initial data set is filled with 1,000 numbers from a minimum value of 1 (for infants age 1) to a maximum value of 99. (People do live to age 100 and beyond, and an infant could technically be 0 years old. However, to keep things simple, assume the lowest age is 1 and the highest age is 99.) With 1,000 data points ranging from 1 to 99, each age in the data set may appear many times. The number of times a particular value appears in a data set is called its *frequency*.

One way to represent this data visually is with a dot plot. Chapter 1 introduced some simple dot plots. In a *dot plot*, the different possible values the data can take are listed along the horizontal axis, and above each value, one dot is placed for every data point in the set that matches that value. Here's a dot plot of the data set of the ages of 1,000 Americans.

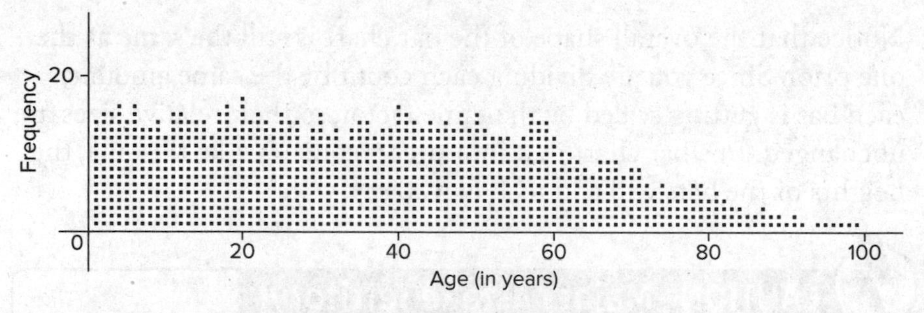

Figure 2–4. Age of 1,000 Americans

The number of dots above each possible data value is equal to the frequency of that value in the set. For example, there are 17 dots above the 20 because age 20 appears 17 times in the data set. Notice that a dot plot shows count data.

You can also present this data as a frequency histogram. A *frequency histogram* is similar to a dot plot except that instead of stacks of dots, bars are used to indicate the frequency of each data value.

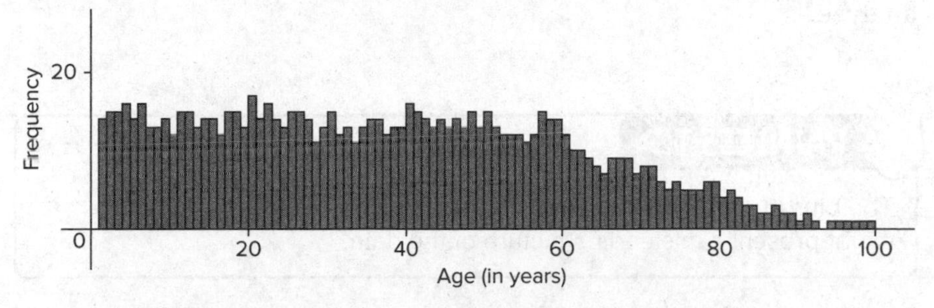

Figure 2–5. Age of 1,000 Americans

Like in a dot plot, the horizontal axis of the histogram lists all the values the data can take and the vertical axis represents the frequency with which each value occurs. In a histogram, the height of the bar indicates the frequency of that value. For example, the bar above 20 has a height of 17, which means the age 20 appears 17 times in the data set. Likewise, the bar for 90 has a height of 1, which means there is only one 90-year-old in the data set.

A histogram looks like a bar chart, but the horizontal axis is a number line, which fixes the order of the bars. You can't rearrange the bars in a histogram the way you can in a bar chart.

Dot plots and histograms are useful in showing how your data is distributed over the possible values it can take. The range of the data is represented along the horizontal axis, and the vertical axis shows how the data is dispersed from the minimum to the maximum value. In the age histogram, you can see that there are more younger people than older people represented in this data set because the bars at the left end are higher than the bars at the right end.

You can modify a histogram by grouping nearby data together into *bins*. For example, when considering the ages of 1,000 Americans, perhaps you are more interested in knowing how many people are in their 20s and 30s and so on, rather than knowing their exact ages. By grouping the data into bins, each bar in the histogram will represent a group of data values instead of a single data value. Choosing an appropriate bin size can make a histogram easier to interpret and analyze.

The bins in a histogram typically have equal widths; otherwise, they might present a misleading picture of the data.

Here's a histogram that shows the age of 1,000 Americans data grouped into bins of width 10.

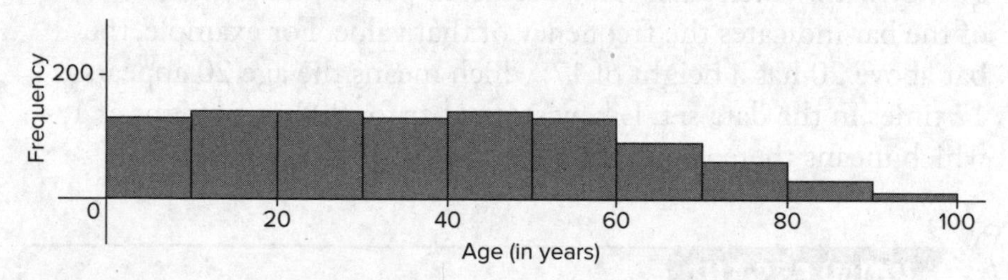

Figure 2–6. Age of 1,000 Americans

The first bar in this histogram represents the first bin of width 10, which includes all the ages greater than or equal to 0 and less than 10. If the variable x represents age, the first bin contains all the ages x where $0 \leq x < 10$. The second bar in the histogram represents the second bin of width 10, which is all the ages $10 \leq x < 20$, the third bin is $20 \leq x < 30$, and so on. Depending on how the bins are defined, you might include the lower limit in the bin and exclude the upper limit, or vice versa. Either method is valid as long as each data point is only counted once. For example, you can't have an age of 20 appearing in both the 10 to 20 bin and the 20 to 30 bin.

As with bar charts, if you add up the heights of the bars in a frequency histogram, you get the total number of data points in the data set. In this example, the height of each bar is the raw count of ages in that bin: 130 for the first bin, 140 for the second bin, and so on. When you add all these up, you get 1,000, the total number of ages in the data set.

Sometimes it's useful to have a histogram that shows the percentages of data that appear in each bin instead of the raw counts. This is known as a *relative frequency histogram*. The height of each bar shows the frequency of the data relative to the size of the entire set. In a relative frequency histogram, the horizontal axis is the same (as that in a histogram that shows raw counts), but the vertical axis starts at zero and goes up to 1 (or 100%).

In a *frequency histogram*, if the heights of the bars show counts, then the sum of the heights of all the bars is equal to the total size of the data set. In a relative frequency histogram, where the heights of the bars show percentages, the sum of the heights will equal 1, or 100%.

Here's a relative frequency histogram of the ages data set.

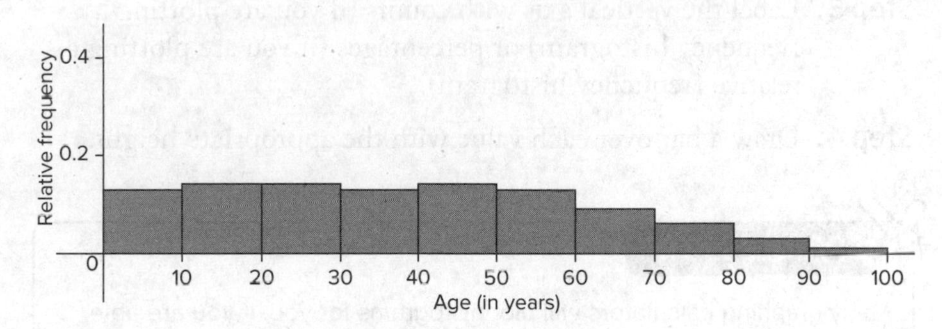

Figure 2–7. Age of 1,000 Americans

In this relative frequency histogram, the first bar has a height of 0.13 because there are 130 ages less than 10 in the data set and $\frac{130}{1,000} = 0.13$. In other words, 13% of the data lies in the first bin. The second bar has height 0.14 because it represents $\frac{140}{1,000} = 0.14$, or 14%, of the overall data, and so on. Relative frequency histograms are useful because they immediately put the data in context. The statement "13% of people are under age 10" is easier to interpret than "130 people are under age 10" because in order to make sense of the second statement, you would need to know how many total people are in the data set.

A percentage, like 54%, can also be written as a pure number, 0.54. That's because "percent" means "out of one hundred." Thus, 54% really means "54 out of 100," which is $\frac{54}{100} = 0.54$. These two representations, frequency histograms and relative frequency histograms, are often used in statistics, so it's important to be comfortable with both.

Now that you know how to analyze frequency histograms and relative frequency histograms, it's important to learn how to create your own histograms. To create a histogram, follow these painless steps.

Step 1: List all the values your data can take along the horizontal axis, and label that axis with the name and unit of the data.

Step 2: Decide how you want to group the data, and choose an appropriate bin size. Determine the frequency for each bin.

Step 3: Label the vertical axis with counts (if you are plotting a frequency histogram) or percentages (if you are plotting a relative frequency histogram).

Step 4: Draw a bar over each value with the appropriate height.

PAINLESS TIP

Many graphing calculators will plot histograms for you. If you are able to enter a list of data into your calculator, you can probably create a histogram by graphing it, so explore your calculator's plotting and statistics capabilities. Remember, everyone uses calculators and computers when working with statistics. Don't be afraid to use yours!

Example 5:

Imagine that a pair of six-sided dice was rolled 1,000 times and the sum of the two dice was recorded each time. To create a histogram of that data, you would first count the number of occurrences of each sum, which you could organize in a table like this.

Sum of Two Dice	Number of Occurrences
2	20
3	60
4	95
5	105
6	150
7	160
8	130
9	115
10	90
11	45
12	30

Now, just follow the painless steps for creating a histogram. This is count data (since you are counting the occurrences of each sum), so in Step 3, use counts. This will result in a frequency histogram, as shown here.

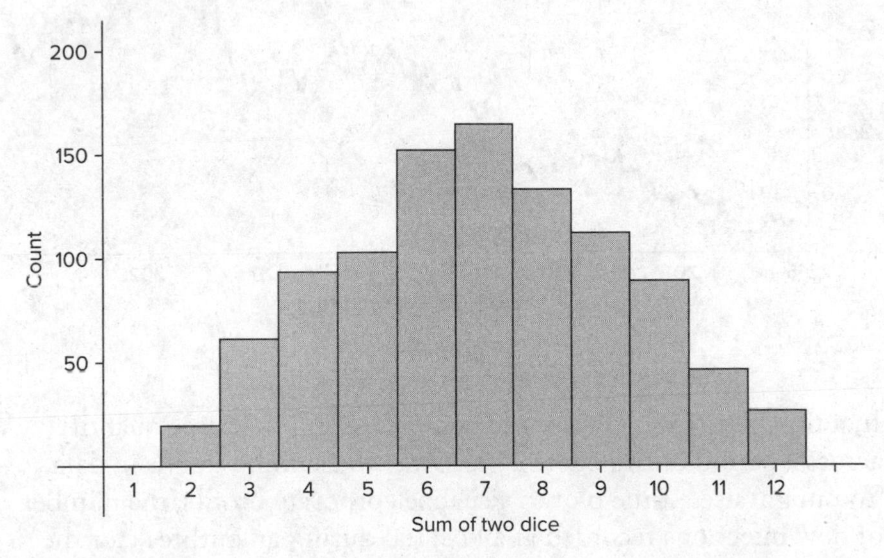

Figure 2–8

This histogram captures and displays many important features of the data. First, notice that the bars of the histogram start at 2 and end at 12 on the horizontal axis. This is because the range of the data is from 2 to 12. (the smallest number on a six-sided die is 1 and the largest is 6, so the minimum possible sum is 2 and the maximum possible sum is 12). You can also see that the bar above the 7 is the highest. This indicates that 7 was the most frequently occurring sum in the 1,000 rolls. Lastly, there is a symmetry present in this histogram. You can imagine drawing a vertical line through the 7, which would split the histogram into two halves that were roughly mirror images of each other. All of these important features of the data are immediately recognizable from the histogram, which shows you how valuable a good visualization of data can be. You'll learn a lot more about interpreting the shape of data in Chapter 4.

Other Common Representations of Data

A *time plot* is a representation that plots quantitative data against a time axis. Time plots are very useful in visualizing how a quantity (for example, a company's stock price) changes over time.

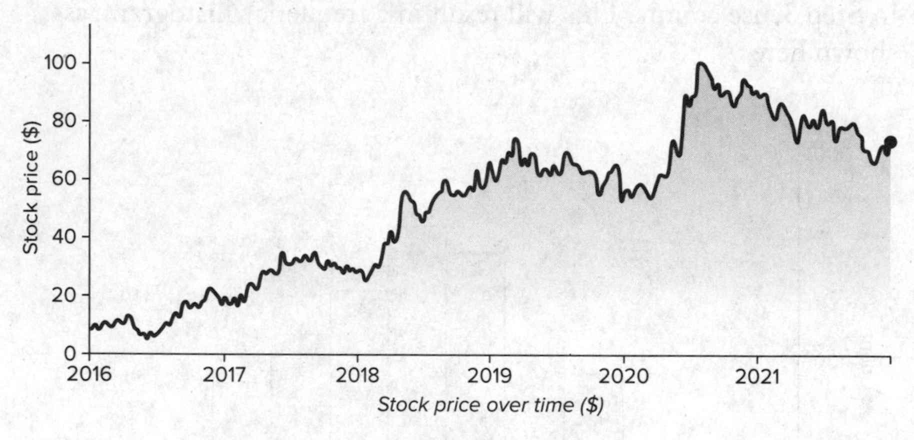

Figure 2–9

In a time plot, the data comes from measuring the same quantity across a period of time. It is a great way to examine trends in data. You might use a time plot to visualize corporate profits, the number of new infections recorded at a hospital during an outbreak, or the number of successful graduates produced by a school district, just to name a few examples.

A *stem plot* (sometimes known as a *stem-and-leaf plot*) is a way to organize small, quantitative data sets. Suppose a biology class earns the following grades on an exam.

75, 58, 90, 97, 92, 82, 66, 78, 90, 85, 91, 67, 71, 82, 89, 68, 75, 67, 73

Biology Exam Scores

The data can be organized in a stem plot as follows.

Biology Exam Scores

Stem	Leaves
9	0 0 1 2 7
8	2 2 5 9
7	1 3 5 5 8
6	6 7 7 8
5	8

Key: 8 | 5 represents 85

Figure 2–10

Here each data value is separated into a "stem" (in this case, the tens digit) and a "leaf" (the ones digit). The first column shows the stems, and the corresponding row shows each leaf associated with that stem. For example, the first row represents the scores of 90, 90, 91, 92, and 97; the second row represents 82, 82, 85, and 89; and so on.

A stem plot is a simple way to organize a small data set that shows you how the data is ordered and distributed. The stem plot orders the data for you and provides information about individual frequencies. Also, the rows of a stem plot are like horizontal versions of the bars of a histogram, showing you where data is clustered.

All the examples of data so far have been *single-variable data*, which occurs when a single kind of information is recorded about an individual (such as the commute time for a person, the age of an adult American, or the price of a stock). This kind of data is described as *univariate*, or "one variable." In contrast, *bivariate* data is information that comes in pairs. Bivariate data is also known as *two-variable data* since each data point is information about two variables. Examples of bivariate data include the height and shoe size for a person, the runs scored and wins for a baseball team, or the population size and gross domestic product for a nation.

A *scatterplot* is a useful way to visualize *bivariate* data. In a scatterplot, each axis represents one of the variables and a point is plotted for each ordered pair of information.

1+2=3 MATH TALK!

An *ordered pair*, often denoted (*x*, *y*), is a single mathematical object that represents two pieces of information. Each piece of information can be thought of as a *coordinate* of the ordered pair.

Example 6:

For seven major American cities, data was collected regarding the size of the population and that city's police budget, as outlined in the following table.

City	Population	Police Budget
Atlanta	0.52 million	$0.25 billion
Chicago	2.7 million	$1.78 billion
Detroit	0.66 million	$0.32 billion
Houston	2.4 million	$0.90 billion
Minneapolis	0.44 million	$0.19 billion
New York City	8.6 million	$11.04 billion
Los Angeles	4.1 million	$1.74 billion

This is bivariate, or two-variable, data because for each individual city, there are two distinct pieces of information: population size and police budget. Each data point can be thought of as an ordered pair. For example, the ordered pair for Atlanta is (0.52 million, $0.25 billion), while the ordered pair for Chicago is (2.7 million, $1.78 billion), and so on. You can then plot each of those points in a scatterplot. In this scatterplot, the horizontal axis represents population and the vertical axis represents police budget.

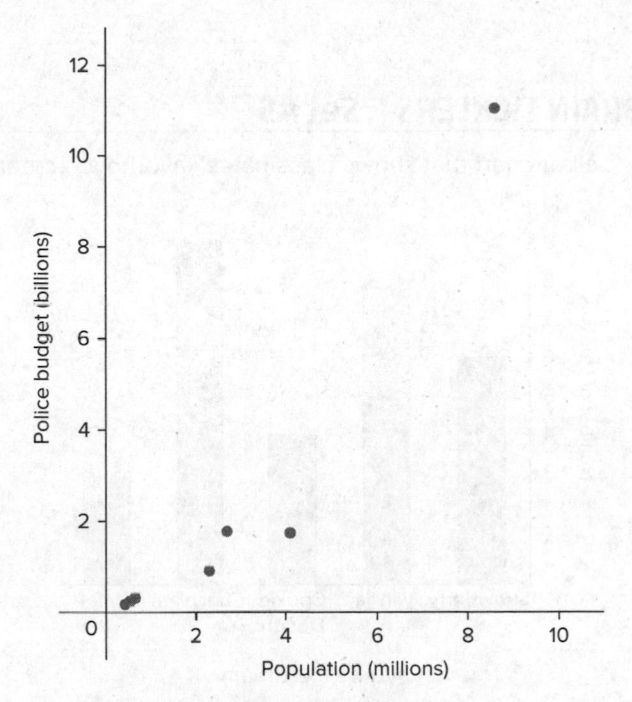

Figure 2–11. Police Budget vs. Population

Scatterplots are useful for seeing trends and relationships in bivariate data. This scatterplot can help you notice a relationship between the population of a city and the size of its police budget. It can also suggest how some cities are similar to each other while others are quite different. A good data visualization makes it easier to ask interesting questions about your data. What questions do you have after seeing this scatterplot? You'll learn about statistical tools that can help make sense of bivariate data later in this book.

BRAIN TICKLERS Set #5

1. Below is a bar chart that shows classmates' favorite ice cream flavors.

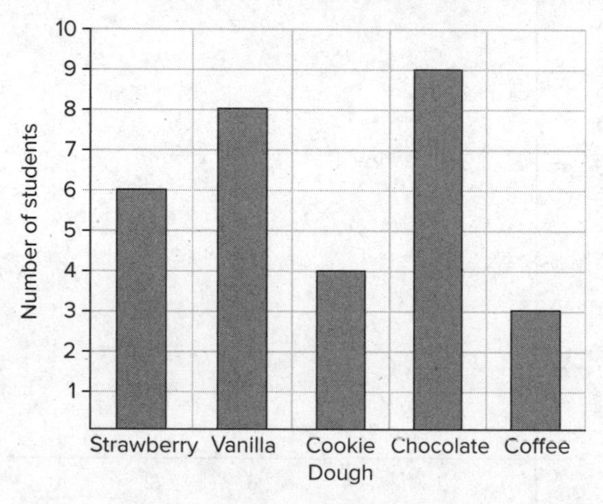

a. Which flavor is most popular?

b. How many people were surveyed for this data?

c. What percentage of people like coffee ice cream best?

2. Here's a relative frequency histogram that shows the results of many rolls of a six-sided die.

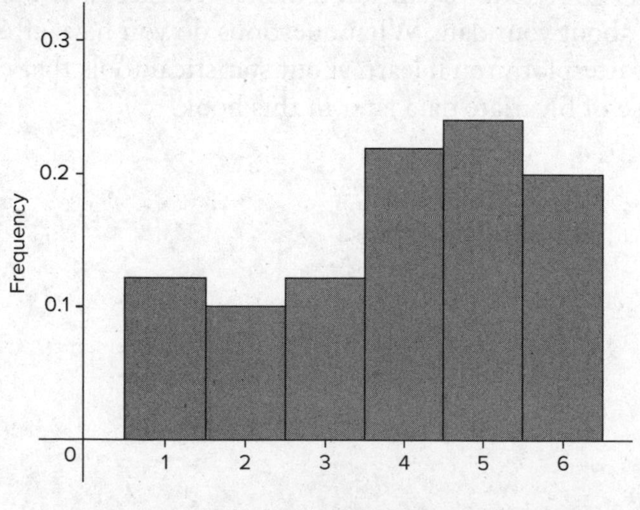

a. Which occurred more frequently: a 1 or a 6?

b. Which occurred more frequently: a die roll less than 4 or a die roll of 4?

c. Explain why it isn't possible to say how many times the die was rolled in collecting this data set.

d. After seeing this histogram, what might you suspect about the die that was rolled?

3. Consider the following data.

49, 32, 20, 29, 10, 11, 31, 34, 16, 20, 14, 32, 27, 20, 14, 23

a. Make a stem plot of this data.

b. What occurs more frequently: numbers in the 10s, 20s, 30s, or 40s?

c. Which individual data value occurs most frequently?

4. Phoenix, Arizona, has a population of 1.7 million people and a police budget of $0.909 billion. Plot this point in the appropriate location on the following scatterplot of police budget vs. population.

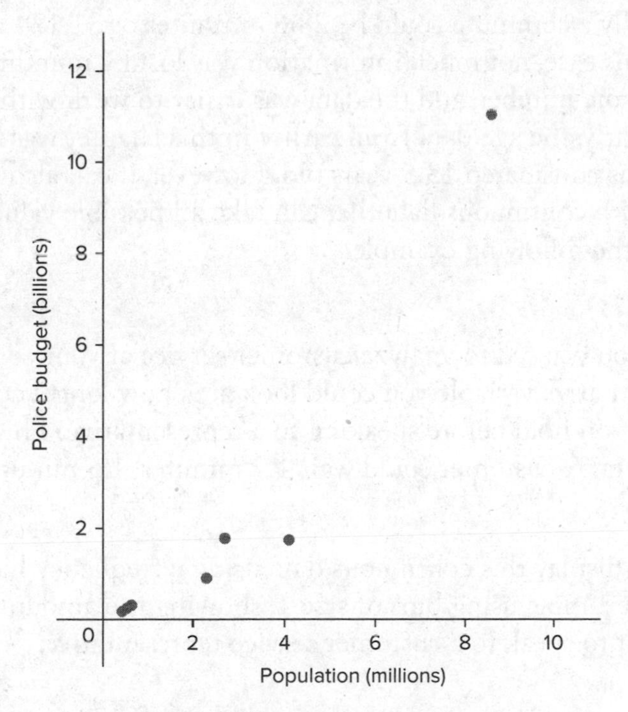

Police Budget vs. Population

(Answers are on pages 41–42.)

Discrete Data vs. Continuous Data

The majority of examples of quantitative data so far have been examples of discrete data. *Discrete data* is data that can only take certain values. For example, the variable "number of siblings" can only take whole number values, like 0, 1, 2, 3, etc. You can't have 2.7 siblings. The number of courses you are enrolled in, the number of states a company does business in, and the number of different stocks owned are all examples of discrete variables.

Continuous data is data that can take any value between whole numbers. How long a patient waits for a hospital bed, the circumference of your skull, or the amount of iron in your blood are all examples of continuous variables. This kind of quantitative data is not restricted to taking whole number values.

Sometimes it is convenient to round continuous data. In Chapter 1, all the commute times were given as whole numbers, even though theoretically a commute could be 28.65 minutes or 31.174 minutes long. In this case, not much information was lost by rounding to the nearest whole number, and the data was easier to work with as a result. Similarly, the age data from earlier in this chapter was rounded. No one was considered 13.5 years old. However, it can also be useful to work with continuous data that can take all possible values, as shown in the following example.

Example 7:

Suppose you wanted to analyze customer service at your company. One quantitative variable you could look at is how long a customer has to wait on hold before speaking to a representative. This is continuous data. A customer could wait 1.2 minutes, 1.5 minutes, or 2.7 minutes.

You could display this continuous data using a frequency histogram. Here's an example using bins of size 1, showing the amount of time callers wait to speak to a customer service representative.

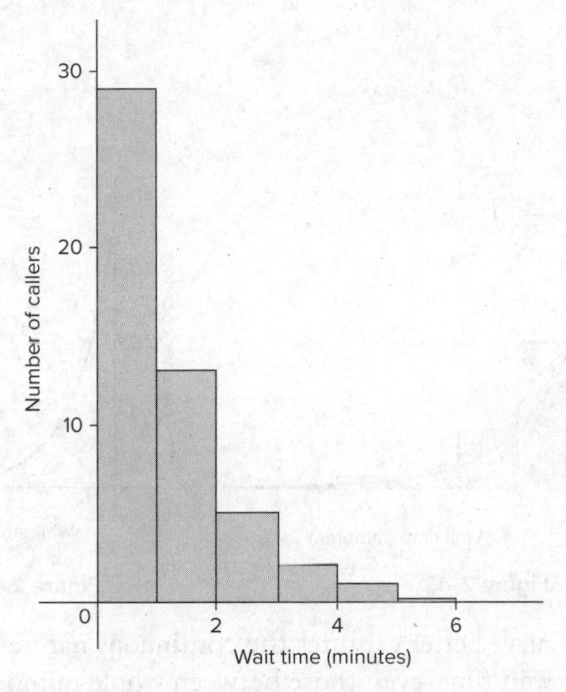

Figure 2–12

Assuming x is the wait time of the caller, the bins here are defined to be $0 \leq x < 1$, $1 \leq x < 2$, and so on. Thus, around 29 people waited less than 1 minute to speak to a representative, 13 people waited at least 1 minute but less than 2 minutes, etc. As with any histogram, if you add up the heights of the bars, you get the total number of data points. In this case, there are 50 callers represented in this data.

Using a histogram in this way almost makes this continuous data appear to be discrete data. The bins in this histogram make it look like the possible data values are whole numbers, such as 1 minute, 2 minutes, etc. However, you can also model or represent continuous data with a smooth curve. Here's an example of a smooth curve modeling this wait time data (first overlaid on the histogram and then shown by itself).

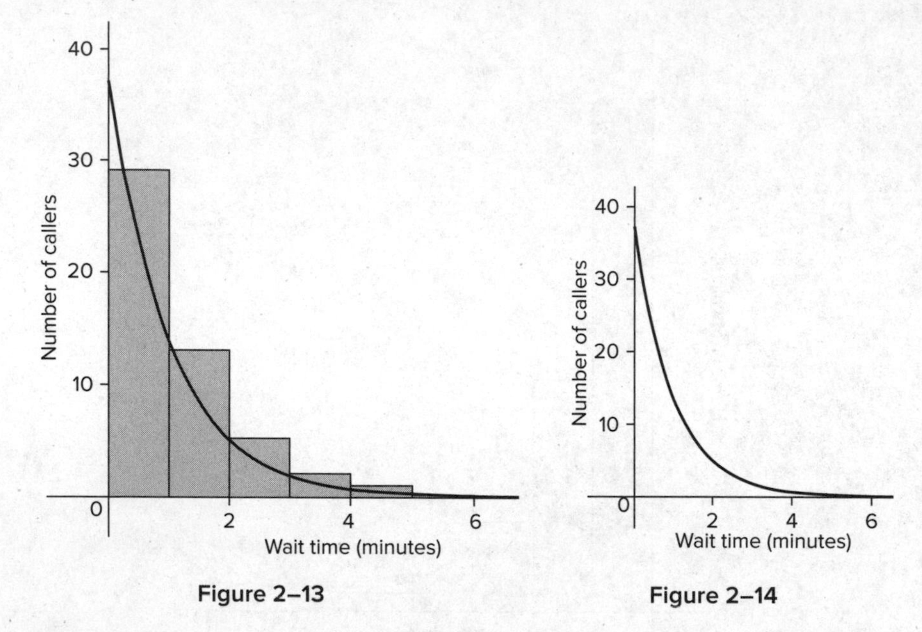

Figure 2–13 **Figure 2–14**

The smooth curve better captures the continuous nature of the data in that every wait time, even those between whole numbers, has a height associated with it. Similar to how in a histogram the height of the bar indicates how many data values are represented, with a smooth curve, the area under the curve represents the amount of data values. Again assuming that x is the wait time for a caller, the area under the curve between $x = 0.5$ and $x = 2.5$ would represent the number of callers who waited between 0.5 minutes and 2.5 minutes. Graphically, that would be represented like this.

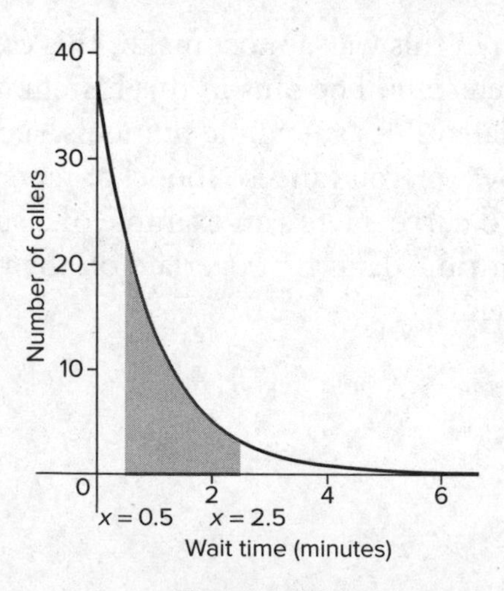

Figure 2–15

Using known mathematical properties of this particular curve, you can calculate this area to be around 19.4, which means that, according to this model of the data, around 19.4 people waited between 0.5 minutes and 2.5 minutes to speak to a representative. Of course, this is an approximation: You can't measure 19.4 people. The smooth curve doesn't match the data exactly, but because it's a close fit, it can be used to model and approximate the data. The virtue of this is that you can use known properties of curves and established mathematical techniques to compute areas, which tell you about the amount of data in certain ranges. You'll see much more of this in Chapter 5 in the discussion of the normal distribution.

The smooth curve in Figure 2–15 shows the distribution of data among its possible values, but another way to represent data is to plot the cumulative distribution. In a *cumulative distribution graph* (also known as a *cumulative distribution function*), the horizontal axis represents the possible data values, the vertical axis represents the cumulative frequency of the data, and the height of the curve indicates the percentage of data less than or equal to that data value.

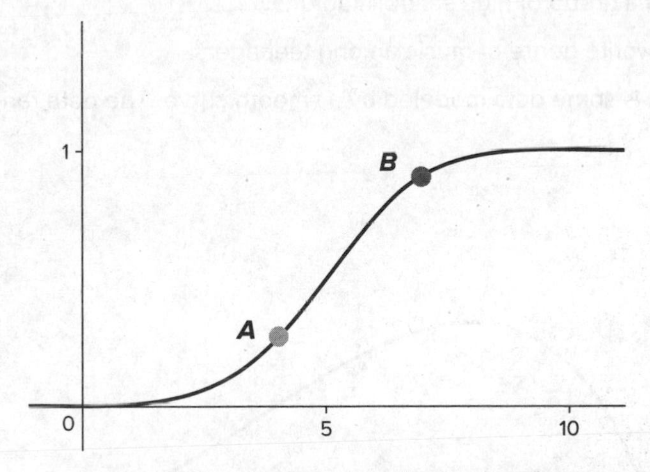

Figure 2–16. Cumulative Distribution Function

In this example, point *A* has coordinates (4, 0.27), which means that 27% of the data is less than or equal to 4. Point *B* has coordinates (7, 0.89), which means that 89% of the data is less than or equal to 7. Notice that as you move to the right in a cumulative distribution, the graph approaches 1 in height because as you move to the right, you include more and more of the data. So eventually you'll reach 100%. Knowledge of cumulative distributions will help you (and your calculator) perform calculations in Chapter 5.

BRAIN TICKLERS Set #6

1. Are the following variables continuous or discrete?
   a. Number of students in a class
   b. Volume of water in a reusable water bottle
   c. Temperature of an oven
   d. Amount of cars parked in a garage
2. Determine which of the following words—quantitative, categorical, univariate, bivariate, continuous, discrete—apply to each of the following data. Note that more than one word can apply to each data.
   a. The number of cars owned by each American family
   b. The amount of sleep per day and the amount of screen time per day for a group of high school students
   c. Favorite genre of music among teenagers
3. Here is some data modeled by a smooth curve. The data ranges from 0 to 5.

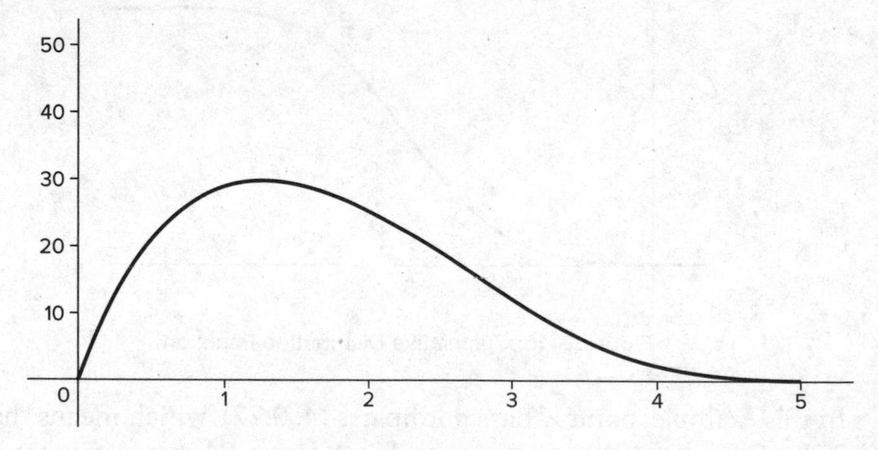

Is more of the data less than 2.5 or greater than 2.5?

(Answers are on page 42.)

Brain Ticklers—The Answers

Set #4, page 20

1. a. Quantitative

   b. Categorical

   c. Quantitative

2. Examples of quantitative variables include yearly revenue (measured in dollars); number of employees (measured in people); and net worth (measured in dollars). Examples of categorical variables include industrial sector (such as technology, food service, or automotive) and primary sales region (Northeast, South, and so on).

3. You might expect higher savings among older Americans, as they have had more time to earn and save money. Of course, there would be exceptions.

Set #5, pages 34–35

1. a. Chocolate

   b. 30

   c. 10%

2. a. 6

   b. A die roll less than 4. The sum of the heights of the three bars for 1, 2, and 3 is greater than the height of the single bar above the 4.

   c. This is a relative frequency histogram. It shows percentages, not counts. So it isn't possible to tell exactly how many times the die was rolled.

   d. The die does not seem to be fair. It seems weighted to produce higher numbers (4, 5, and 6) more frequently than lower numbers (1, 2, and 3).

3. a.

Stem	Leaves
4	9
3	1 2 2 4
2	0 0 0 3 7 9
1	0 1 4 4 6

Key: 4 | 9 represents 49

b. Numbers in the 20s occur most frequently.

c. 20

4.

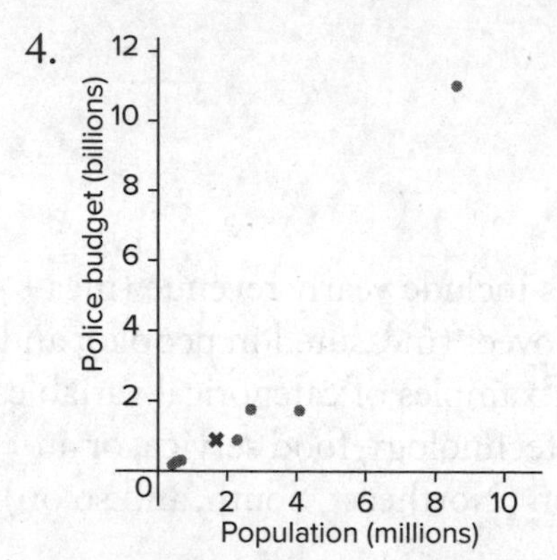

Police Budget vs. Population

Set #6, page 40

1. a. Discrete

   b. Continuous

   c. Continuous

   d. Discrete

2. a. Quantitative, univariate, discrete

   b. Quantitative, bivariate, continuous (both variables)

   c. Categorical, univariate

3. Less than 2.5. The amount of data between 0 and 2.5 is represented by the area of this region under the curve.

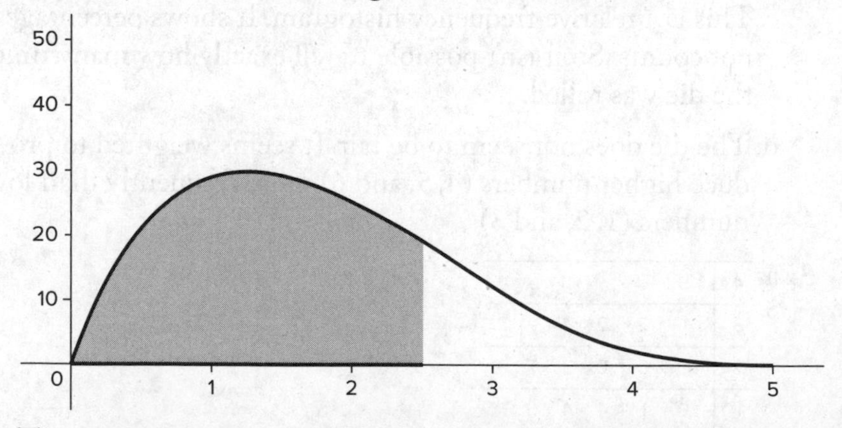

The area of this region appears to be much greater than the area of the region between 2.5 and 5.

Chapter 3

Descriptive Statistics

As you saw in Chapter 1, statistics can help you understand a data set by capturing its important features in a few simple numbers called summary statistics. These are also known as *descriptive statistics* because they describe the data. In this chapter, you'll learn how simple descriptive statistics can be used to understand the key characteristics of quantitative data: center and spread.

Measures of Center

The center is an important feature of a data set. The center gives you an idea of what "average" data looks like. Since data is often clustered around its center, and even symmetric about its center, knowing about the center is a good first step toward understanding the data as a whole.

There are three descriptive statistics that help you think about the center of a data set: *mean*, *median*, and *mode*. Each of these offers a slightly different *measure of central tendency* of a set of data.

Mean

As you saw in Chapter 1, the mean of a data set is the sum of all the values in the set divided by the number of values in the set. The mean is also frequently referred to as the *average*, and it can be thought of as an "average" member of the data set. Here are a couple of quick examples to remind you how to calculate the mean of a data set.

Example 1:

Suppose a student asked each of her classmates how many siblings they have and recorded their answers.

$$1, 0, 1, 2, 5, 1, 0, 2, 3, 1, 2, 0$$

Number of Siblings Among Classmates

To compute the mean of this data set, you would sum the twelve values and then divide that sum by 12:

$$\frac{1+0+1+2+5+1+0+2+3+1+2+0}{12} = \frac{18}{12} = 1.5$$

The mean of this data set is 1.5 siblings, so in a sense, the "average" person in class has 1.5 siblings. Of course, no one has 1.5 siblings: "number of siblings" is a discrete variable and can only take integer values.

The mean doesn't have to be a member of the data set.

Example 2:

A week's worth of stock prices for Company A's stock are recorded.

$34.62, $43.09, $40.78, $37.24, $36.99

Stock Prices for Company A

The mean of this data is:

$$\frac{34.62 + 43.09 + 40.78 + 37.24 + 36.99}{5} = 38.544$$

Since stock price is usually reported to the nearest cent, you would round the result and report that the mean stock price this week was $38.54. Again, you can see how the mean gives you a sense of the average behavior of a data set. Throughout the week, the stock price rose and fell, but on average it was around $38.54.

The unit of the mean is the same as the unit of the data itself.

For a data set that contains n values of a variable x, the mean can be expressed by the following formula:

$$\frac{1}{n}\sum x$$

The expression $\sum x$ is read "sigma x" or "the sum of all x."

In math and statistics, the Greek capital letter $\sum$ is an instruction to add up a set of numbers, so $\sum x$ is the sum of all the values of x, that is, all the values in the data set. Since n is the size of the set and multiplying by $\frac{1}{n}$ is the same as dividing by n, this formula gives you the mean: the sum of all the values divided by the number of values.

Means are often denoted using *bar notation*. The mean of the data set can be written as $\overline{x}$ (which is read as "x bar"), so you might also see the formula written this way:

$$\overline{x} = \frac{1}{n}\sum x$$

REMINDER

This kind of mean is also known as the *arithmetic mean* because it involves adding up numbers. This name distinguishes it from other kinds of means in mathematics that involve other operations, like multiplying and dividing.

Median

The mean is the most common measure of center, but the *median* provides another important measure of center of a data set. While the mean can be thought of as an "average" data value, the median can be thought of as the middle data value. If you order your data from least to greatest, the median will be the number in the middle.

Example 3:

Imagine that you want to find the median times for the commute data (which you first saw in Chapter 1). Here's the data from Commute B, arranged in order from least to greatest.

$$34, 35, 36, 36, 37, 37, 37, 37, 39, 39, 40, 40, 41, 41, 41$$

Commute B Times

To find the median, simply locate the middle number. Since there are 15 numbers in this data set ($n = 15$), the middle number is the 8th number in the list, which is 37.

$$34, 35, 36, 36, 37, 37, 37, \underset{\substack{\wedge \\ \text{median}}}{37}, 39, 39, 40, 40, 41, 41, 41$$

Now if you try the same thing with the data from Commute A, you run into a problem.

$$26, 27, 27, 30, 30, 31, 32, 42, 42, 43, 44, 46$$

Commute A Times

There are an even number of values in this data set ($n = 12$), so there is no number exactly in the middle. When this happens, the median is defined as the mean of the two numbers closest to the middle. In this case, the median is the mean of 31 and 32 (the 6th and 7th numbers on the ordered list), which is $\frac{31 + 32}{2} = 31.5$. Thus, 31.5 minutes is the median of this data set.

PAINLESS TIP

Just as with the mean, the median doesn't have to be a member of the data set. In this example, the values in the data set are all whole numbers, but the median is not a whole number.

Here's a summary of the painless steps for finding the median of a data set.

Step 1: Rearrange the data set in order from least to greatest.

Step 2: Determine whether n is odd or even. If n is odd, find the number in the middle of this ordered list, and that is the median.

Step 3: If n is even, there is no single middle number. So calculate the mean of the two middle values, and that is the median.

The median measures the center of a data set in a different way than the mean. As the middle number, the median basically divides the data into two parts of equal size: one of those parts lies below the median and the other part lies above the median. This gives the median a very important property when it comes to understanding data. In a data set, approximately 50% of the data is below the median and 50% of the data is above the median.

Like the mean, the median can also serve as an average representative of a data set. If you research home prices in your area, you might find that when the "average" price of a home is mentioned, it's the median, not the mean, that is being discussed. This is because the median can be a more useful measure of center for home price data.

Example 4:

Imagine a neighborhood with 100 homes: 90 of which cost \$100,000, and 10 of which cost \$2,000,000. The median home price is \$100,000, but the mean home price is \$290,000. In this neighborhood, the vast majority of houses cost \$100,000, so the median is much more representative of the cost of an "average" house than the mean.

CAUTION—Major Mistake Territory!

Be careful when using the word "average"; it can mean different things in different contexts. Usually when people say "average," they are talking about the mean, but sometimes they are referring to the median. When you encounter the word "average," make sure you know exactly what it refers to.

In this example neighborhood, the handful of very expensive houses has a big impact on the mean home price, pushing it up from $100,000 to nearly $300,000. This is an example of how the mean can be sensitive to large data values. Here's another simple example that shows how the mean can be sensitive to large data values while the median can be much less sensitive to these large values.

Example 5:

Consider this small list of numbers.

$$1, 2, 3, 4, 5, 6, 7, 8, 9$$

The mean and median of this list are both 5. Now, add a very large number to the list.

$$1, 2, 3, 4, 5, 6, 7, 8, 9, 1000$$

Adding this new number changes the mean from 5 to 104.5, a dramatic increase. The median, however, only changes from 5 to 5.5. The median is much less sensitive to extreme data values than the mean. Put technically, the median is *resistant* to extreme observations, unlike the mean, which is *nonresistant*. The fictional neighborhood from Example 4 is another example of the resistance of the median as a measure of central tendency. The few very expensive houses drove the mean price up but didn't have an outsized impact on the median.

Mean and median form a useful one-two punch when it comes to measuring the center of a data set. Often, the mean and median of a data set will be close to one another, which tells you something useful about the data. When they are not close together, that tells you something important too. You'll learn more about this in Chapter 4 when you study different distributions of data.

Mode

The *mode* is the third "m"-word in measures of central tendency. The mode is the most frequently occurring member of a data set.

Example 6:

You have two possible commuting methods (Commute A and Commute B). From the data recorded for each commute, you would like to know the most frequently occurring time. Here's the Commute B data.

$$34, 35, 36, 36, 37, 37, 37, 37, 39, 39, 40, 40, 41, 41, 41$$

In this data, the mode is 37. It occurs four times in the data, and nothing occurs more frequently. Now, here's the data set for Commute A.

$$26, 27, 27, 30, 30, 31, 32, 42, 42, 43, 44, 46$$

In this data set, there are three modes: 27, 30, and 42 all occur twice, and nothing occurs more frequently.

As with the median, finding the mode of a data set is easier when you arrange the data in order. Some representations of data make finding the mode easy, as seen in the following two examples.

Example 7:

You have the same commuting data from Example 6, but this time, you'd like to visualize the data in dot plots and use those visualizations to determine the mode. The dot plots contain information about the frequency of the data, which is exactly what the mode conveys to you.

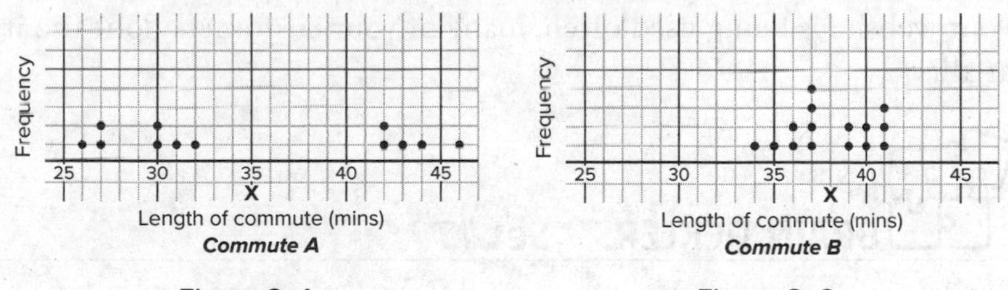

Figure 3–1

Figure 3–2

In the dot plot on the right, the tallest stack of dots sits above the 37, which is the mode of the data set from Commute B. On the left, there are three stacks of 2 dots (above 27, 30, and 42), corresponding to the three modes of the data from Commute A.

Example 8:

A teacher has the following data set of students' biology exam scores and would like to determine the mode of these scores.

$$75, 58, 90, 97, 92, 82, 66, 78, 90, 85, 91, 67, 71, 82, 89, 68, 75, 67, 73$$

At first glance, it's not immediately obvious which value is most frequent. However, a stem plot can make the mode much easier to find.

Biology Exam Scores

9	0 0 1 2 7
8	2 2 5 9
7	1 3 5 5 8
6	6 7 7 8
5	8

Key: 8 | 5 represents 85

Figure 3–3

When the data is organized in a stem plot, it's easy to see that 67, 75, 82, and 90 each appear twice and no other value appears more often.

Mode is less often used as a measure of central tendency for data sets, but it can convey useful information about the nature of data. If you run a company and see that existing customers, on average, visit your website 5.4 times per month, knowing that the mode of that data is 0 would be very helpful. It would tell you that even though your website is being used often, many of your customers don't use it at all.

BRAIN TICKLERS Set #7

1. Find the mean, median, and mode of the following data sets:

   a. 100, 200, 300, 400, 500, 600, 700, 800, 900, 1000

   b. 8, 3, 7, 2, 3, 4, 6, 3, 108, 2, 6, 3, 1

   c. 7, 7, 7, 7, 7, 7, 7, 7

2. Create a data set of positive numbers with $n = 5$ that has a mean of 10 and a median of 3.

3. The mean of the data set $\{a, b, c, d, e\}$ is 10, and the median is 6.

   a. What is the mean of the data set $\{a + 3, b + 3, c + 3, d + 3, e + 3\}$?

   b. What is the median of the data set $\{a + 3, b + 3, c + 3, d + 3, e + 3\}$?

   c. What is the mean of the data set $\{2a, 2b, 2c, 2d, 2e\}$?

   d. What is the median of the data set $\{2a, 2b, 2c, 2d, 2e\}$?

4. The mean of the data set {*a*, *b*, *c*, *d*, *e*} is 10. The mean of the data set {*a*, *b*, *c*, *d*, *e*, *f*} is 20. What is *f*?
5. Consider the following histogram.

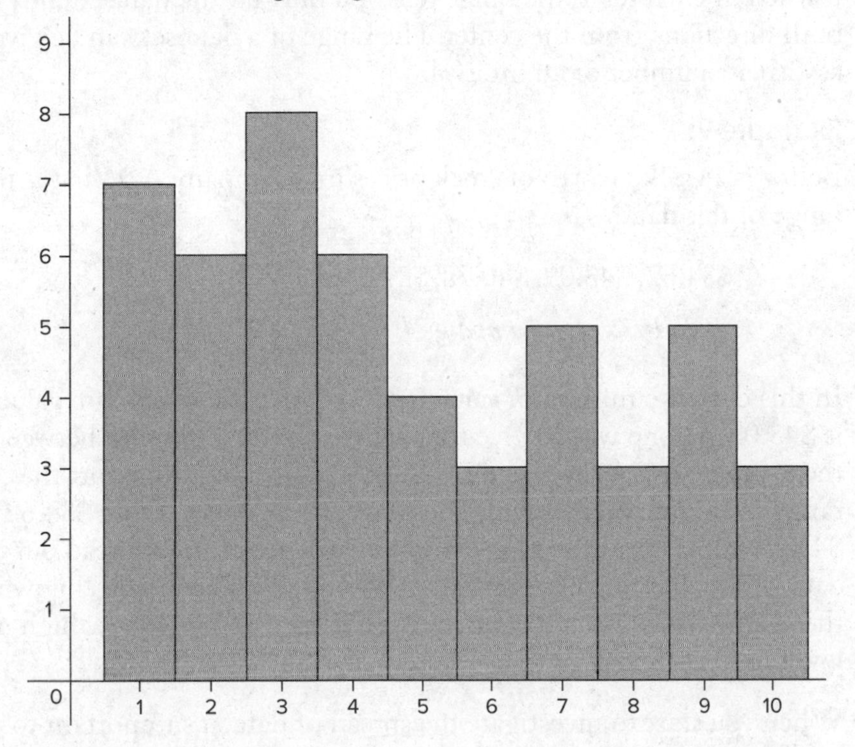

a. What is the mode of this data set?

b. What is the size of the data set?

c. How many elements in this data set are less than or equal to 2?

(Answers are on page 68.)

Measures of Spread

Once you understand the center of quantitative data, you next want to understand the spread. The *spread*, or *dispersion*, of data is how the data is distributed over its different possible values. Where is the data clustered or concentrated? How far does it extend? Together with the center, knowing the spread helps you better understand the data as a whole.

Range

The simplest measure of the spread of data is the *range*. As you learned in Chapter 1, the range tells you how far the data extends in both directions from the center. The range of a data set can be given as either a number or an interval.

Example 9:

Below is a week's worth of stock prices for Company A. What is the range of this data?

$34.62, $43.09, $40.78, $37.24, $36.99

Stock Prices for Company A

In this data, the minimum value is $34.62 and the maximum value is $43.09. So one way to give the range is as the difference between these values: $43.09 – $34.62 = $8.47. You could also report the range as an interval by saying the range of the data is from $34.62 to $43.09. Either way, the range gives you a sense of how spread out the data is, which is useful information. When you're investing, knowing the average price is important, but you'll want to know how high or low the stock prices go too.

When you start to investigate the spread of data, it's important to consider potential outliers. An *outlier* is a data point that does not fit in with the overall pattern of the data. When you spot an outlier, you should ask yourself whether that particular data point really belongs with the rest of the data set or if there's some reason you might wish to exclude it from your analysis.

For example, suppose you recorded a second week's worth of stock prices.

$38.47, $41.12, $405.70, $39.96, $39.92

Stock Prices for Company A

That third data point looks very unusual; it doesn't seem to fit with the rest of the data. It's an outlier, so before you start performing any statistical analysis, you should take a closer look. It's possible that the company's stock rose from $41 to $405 in one day, but that's very unlikely. It's far more likely that there was a transcription error.

Perhaps when copying down the stock price, someone wrote $405.70 instead of $40.57. If that's the case, you don't want to include $405.70 in your data set. That extreme data value will dramatically affect descriptive statistics like the mean and the range and render your analysis useless. Instead, you would want to discard this outlier from the data before moving on. However, outliers aren't always the result of a data collection error, as seen in the following example.

Example 10:

Suppose that one day while recording data about your daily commute, your train breaks down and you end up stuck on the track for an hour. As a result, your commute time for that day ends up being 100 minutes, which looks very unusual among the rest of the data.

34, 35, 36, 36, 37, 37, 37, 37, 39, 39, 40, 40, 41, 41, 41, 100

Commute B Data with Outlier

This 100-minute commute wasn't an error in data collection; it is an actual commute time. However, because the circumstances of that commute were so unusual (a train malfunction), you might not wish to include it in your analysis. It isn't really representative of what the commute is like, and it will dramatically alter statistics like the mean and range. On the other hand, if on subsequent commutes you discover that such a delay isn't that unusual, then you might want to consider that data point as part of your analysis.

When you spot an outlier, whether you end up discarding or keeping it depends on the nature of your data.

PAINLESS TIP

There's no clear-cut way to handle an outlier. The key is to identify outliers and ask yourself if they really belong in the data set. Often, the best way to answer that question is to try to collect more data and see just how unusual the outlier is.

Quartiles

As you saw earlier in this chapter, the median of a data set splits the data into an upper half and a lower half. Here's the Commute B data from Example 3, with the median splitting the data set in half.

34, 35, 36, 36, 37, 37, 37, 37, 39, 39, 40, 40, 41, 41, 41
^
median

You can repeat that process and split each half into two halves by finding the upper quartile and the lower quartile.

34, 35, 36, 36, 37, 37, 37, 37, 39, 39, 40, 40, 41, 41, 41
^ ^ ^
1st Q 2nd Q 3rd Q

The *lower quartile* (or *1st quartile*) is the median of the lower half of the data. The *upper quartile* (or *3rd quartile*) is the median of the upper half of the data. Together with the *median* (also known as the *2nd quartile*), the quartiles split the data up into four equal-size parts or quarters. One quarter of the data, or 25%, lies below the 1st quartile, and 25% of the data lies above the 3rd quartile. Notice that since the 2nd quartile is the median, it makes sense that 50% of the data lies below and 50% lies above it.

Here's a summary of the painless steps for finding the quartiles of a data set.

Step 1: Order the data from least to greatest.

Step 2: Find the middle number. This is the median of the data set (also known as the 2nd quartile).

Step 3: Find the middle number of the data below the median. This is the 1st quartile.

Step 4: Find the middle number of the data above the median. This is the 3rd quartile.

When finding quartiles, if there is no middle number, take the mean of the two numbers closest to the middle, just as you would when finding the median.

Once you have the quartiles of a data set, you can compute the interquartile range. This is another simple statistic that captures the spread of data. *Interquartile range* (IQR) is the distance between the first and third quartile and can be computed using the following formula:

$$IQR = Q_3 - Q_1$$

Q_3 (read "Q sub 3") is the third quartile, and Q_1 ("Q sub 1") is the first quartile. In the case of the Commute B data, the interquartile range is $40 - 36 = 4$ minutes. This tells you that 50% of the data is clustered in a range of 4 minutes around the center (here, the median) of the data set.

The interquartile range can also be used to test a data point to determine whether it is an outlier. Here are the painless steps for that test.

Step 1: Compute the interquartile range (IQR) of your data.

Step 2: Multiply the IQR by 1.5.

Step 3: If your data point is more than $1.5 \times$ IQR below the first quartile or more than $1.5 \times$ IQR above the third quartile, then by this criterion, declare the data point a statistical outlier.

You can use this test to establish that a 100-minute commute for Commute B is an outlier. Step 1 was already completed above, determining that the IQR is 4 minutes and the upper quartile is 40 minutes. As per step 2, multiply the IQR by 1.5:

$$1.5 \times \text{IQR} = 1.5 \times 4 = 6$$

Finally, following step 3, you see that $Q_3 + 6 = 40 + 6 = 46$. Since $100 > 46$, you know that the 100-minute commute is considered an outlier by this test.

The idea behind this outlier test is that half of the data lies in a region between the 1st and 3rd quartiles. So being far away from that region is an indication that a particular data value is extreme when compared to the rest of the data, for example, a 100-minute commute. This test involving IQR gives you one way to quantify what "extreme" means in this context and can inform your decision about whether or not to include that data point in your analysis.

There's no single procedure for identifying and dealing with outliers. Just remember that when you spot some data that looks out of place, you should investigate it more fully to determine whether it's data you should use or exclude.

With the quartiles, you can give the *five-number summary* of the data. This is a simple description of a data set using just five numbers: the minimum, the lower quartile, the median, the upper quartile, and the maximum. Here's the five-number summary of the data from Commute B.

Min	Q1	Med	Q3	Max
34	36	37	40	41

You can tell a lot about the data set from just these five numbers. The median tells you a measure of center. The minimum and the maximum give you the range. The properties of the quartiles tell you about the distribution of the data: roughly half the data lies between 36 and 40, 25% of the data is above 40, and 25% is below 36. You can also turn the five-number summary into a visualization of the data with a box plot.

Box Plots

A *box plot* (sometimes called a *box-and-whisker plot*) captures all five elements of the five-number summary: the minimum, the lower quartile, the median, the upper quartile, and the maximum.

Example 11:

1,000 Americans (ages 1 to 99) are randomly selected, and a data set of their ages (in years) is created. Here's the five-number summary of that data.

Min	Q1	Med	Q3	Max
1	18	36	55	99

To create a box plot for a data set like this, follow these painless steps.

Step 1: Determine the five-number summary for your data set. That's the minimum, the lower quartile, the median, the upper quartile, and the maximum. (In this example, that summary has already been supplied to you.)

Step 2: Draw a rectangle whose ends are the lower quartile and the upper quartile. Then draw a line inside the rectangle above the median.

Step 3: Draw the "whiskers" from the rectangle that extend to the minimum and maximum of the data set.

Here's the corresponding box plot.

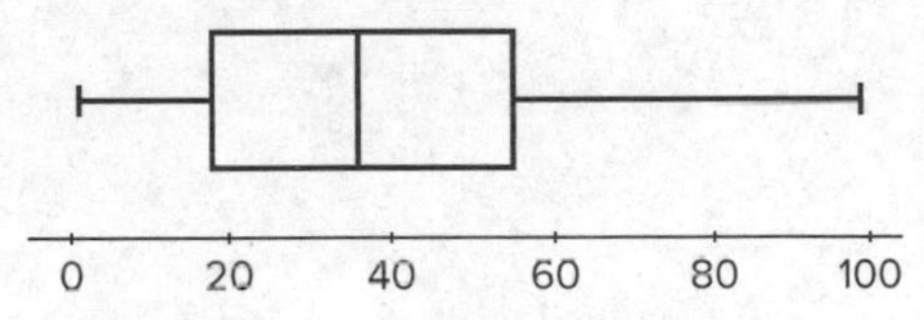

Figure 3–4. Ages of 1,000 Americans

The rectangle in the middle is the "box." The sides of the box are the lower quartile (18) and the upper quartile (55), and the line inside the box represents the median (36). The ends of the plot (the "whiskers") are the minimum (1) and the maximum (99).

PAINLESS TIP

When drawing a box plot, make sure to draw it to scale, as the size and position of the box indicate specific statistical information about the data. Also, don't assume the median automatically goes in the middle of the rectangle.

Box plots provide a simple picture of how a data set is distributed. Because of the way quartiles work, you know that 50% of the data lies in the "box" and 25% lies in each "whisker" or "tail." Looking at the box plot of age data, you can see that the ages are not evenly spread out. The box is shifted to the left, and since the box contains half the data, this means the majority of the data lies to the left. Since the longer whisker, or tail, extends to the right, this shows you that the upper quartile (the top 75% of the data) is much more spread out than the lower quartile (or the bottom 25%).

Percentiles

Quartiles divide a data set into four parts of equal size. You can divide a data set more generally using *percentiles*. A common use of percentiles is in describing test data. Here's a smooth curve that models the scores from an aptitude test.

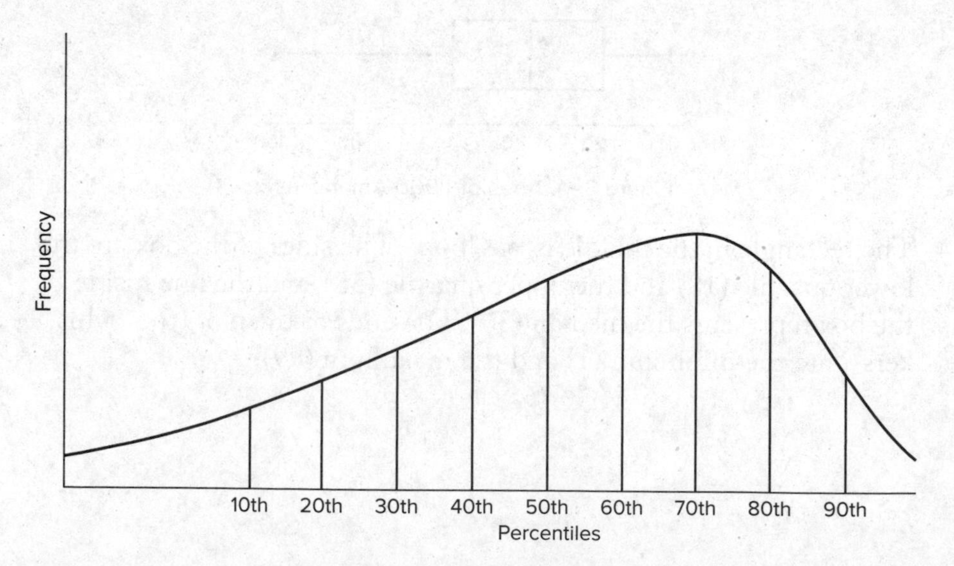

Figure 3–5. Test Scores

Recall that when data is modeled by a smooth curve, the amount of data in any range is given by the area under the curve. Here the data has been divided into 10 regions of equal area, which means that each region contains 10% of the data.

The marks along the horizontal axis indicate percentiles. The leftmost percentile marked on the axis is the 10th percentile, so the region to the left of this mark contains 10% of the data. Next over is the 20th percentile mark, so 20% of all the data lies to the left of this mark (which also means 10% of the data lies between the 10th and 20th percentiles). The rightmost percentile on the axis is the 90th percentile. This means 90% of the data lies to the left of this marker and 10% lies to the right. So, if you score in the "90th percentile" on a test, that means your score is higher than 90% of all the scores. A score in the 80th percentile is higher than roughly 80% of the data, and so on. Raw scores can be converted to percentiles to make interpretation and comparison of data easier. Knowing your test score is in the 90th percentile means it is higher than 90% of all test scores, which might be more useful information than knowing that your raw test score was, say, 73 out of 105.

In the same way that quartiles divide a data set into four equal parts, percentiles can be used to divide the data into any number of parts of known size. This can be helpful in analyzing and exploring how your data is distributed.

BRAIN TICKLERS Set #8

1. Consider the following data set.

   1, 10, 7, –5, 0, –3, 2, –1, 0, 9, 4, 5, 8, 6, 3

   a. Find the five-number summary for this data set.

   b. Sketch a box plot of this data.

   c. Compute the interquartile range. Is –5 an outlier according to the IQR test?

2. What percentiles are equivalent to the upper and lower quartiles?

3. Suppose Company X's stock price is currently $1,000 per share and Company Y's stock price is currently $10 per share. Which data set do

you suspect would have the larger range: Company X's stock price over the course of a week or Company Y's stock price over the course of the same week?

(Answers are on pages 68–69.)

Deviation

The most useful measure of the spread of data is deviation. As you saw in Chapter 1, deviation is a measure of how far data is from the mean of the data set. Each value in a data set has its own individual deviation from the mean, but you can also compute various "average" deviations for an entire data set.

Example 12:

Recall the stock price data for Company A.

$34.62, $43.09, $40.78, $37.24, $36.99

Stock Prices for Company A

This data has a mean of $38.54, so the individual deviations are found by subtracting the mean from each data point:

(34.62 – 38.54), (43.09 – 38.54), (40.78 – 38.54),
(37.24 – 38.54), (36.99 – 38.54)

–$3.92, $4.55, $2.24, –$1.30, –$1.55

Deviations of Stock Prices

Since you are subtracting dollars from dollars, the unit of deviation is dollars. In general, the unit of deviation will be the same as the unit of the data. Also, notice that some of these numbers are negative. That's because some stock prices are below the mean stock price, which is being subtracted from each value. As mentioned in Chapter 1, you usually want deviations to be positive. So you can just take the absolute value of each deviation, which turns any negative numbers into positive numbers.

$3.92, $4.55, $2.24, $1.30, $1.55

Deviations of Stock Prices

This list shows the individual deviations from the mean for each data point in the set, and you can use this to compute an average deviation across the entire data set. Here are two ways to do that.

Mean Absolute Deviation

One way to measure the average deviation from the mean in a data set is to simply take the average of all the individual deviations. This is known as the *mean absolute deviation* (or just *mean deviation*), and it was covered in more depth in Chapter 1. Here's a quick refresher on the painless steps for computing the mean absolute deviation (MAD).

Step 1: Compute the mean of the data set.

Step 2: Calculate the absolute value of each individual data point's deviation from the mean. This will form a new data set.

Step 3: Compute the mean of this new data set of deviations.

The technical formula for mean absolute deviation is:

$$\text{MAD} = \frac{1}{n}\sum|\overline{x} - x|$$

This formula might look complicated, but it just captures everything mentioned in the painless steps. Remember that $\overline{x}$ is just the mean of the data. So for each data value, x, the quantity $\overline{x} - x$ is an individual deviation from the mean, and taking the absolute value guarantees it's positive. The $\sum$ tells you to add up all those individual deviations, and the $\frac{1}{n}$ is there to divide by n to give you the average deviation.

In the case of the stock prices from Example 12, the mean absolute deviation is:

$$\frac{\$3.92 + \$4.55 + \$2.24 + \$1.30 + \$1.55}{5}$$

This is approximately \$2.71 and is simply the average deviation from the mean; thus, on average, each daily stock price differs from the mean stock price by around \$2.71. Notice that since each deviation is measured in dollars, the average deviation is also measured in dollars.

Standard Deviation

Although mean deviation is a simple and useful way to measure the deviation of a data set, the *standard deviation* is far more commonly used in statistics as a measure of the dispersion, or spread, of data. The purpose of standard deviation is the same as that of mean deviation: to measure how far, on average, data is from the mean. However, standard deviation takes a slightly different approach. Instead of averaging the individual deviations, you average the square of the individual deviations and then take the square root of that average.

1+2=3 MATH TALK!

"Squaring" a number means multiplying it by itself. This comes from the fact that the area of a square is the side length times itself: for example, the area of a square with side length 5 is 5 squared. That is, $5^2 = 5 \times 5 = 25$.

Here are the painless steps for computing the standard deviation of a data set.

Step 1: Compute the mean of the data set.

Step 2: Form a new data set by computing all the individual deviations from the mean.

Step 3: Square each number in this new data set.

Step 4: Compute the mean of this new data set.

Step 5: Take the square root of the mean you found in Step 4. This is the standard deviation.

Here's an example of these steps put into action.

Example 13:

Here is the stock price data for Company A from Example 12.

$34.62, $43.09, $40.78, $37.24, $36.99

Stock Prices for Company A

The mean (Step 1) is \$38.54, and the individual deviations of the stock prices (Step 2) are as follows:

\$3.92, \$4.55, \$2.24, \$1.30, \$1.55

Deviations of Stock Prices

Now, find the average of the square of each deviation (Steps 3 and 4):

$$\frac{3.92^2 + 4.55^2 + 2.24^2 + 1.30^2 + 1.55^2}{5} = 9.04$$

Finally, take the square root of that average to get the standard deviation (Step 5):

$$\sqrt{9.04} = 3.01$$

A table can help you organize standard deviation calculations. Here's a table that displays each data value, x, its individual deviation, $\bar{x} - x$, and the square of the individual deviation $(\bar{x} - x)^2$.

x	$\bar{x} - x$	$(\bar{x} - x)^2$
34.62	−3.92	15.37
43.09	4.55	20.70
40.78	2.24	5.02
37.24	−1.30	1.69
36.99	−1.55	2.40

Then to compute the standard deviation, you just add up the last column, divide by 5, and take the square root of that result.

The unit of standard deviation is still the same as the unit of the original data. Early in the process of computing standard deviation, the stock prices were squared, producing a unit of "dollars squared." By taking the square root at the end of the process, however, the unit was converted back into dollars. So, the standard deviation of these stock prices is \$3.01.

The fact that the mean and standard deviation are measured in the same units as the data itself makes it easier to use them together to analyze and interpret data. It can also help you make sure your answers make sense in the context of the problem.

The standard deviation of these stock prices, \$3.01, is similar to the mean deviation, \$2.71, but is a bit higher. It is generally true that the standard deviation of a data set is higher than the mean deviation; that is because standard deviation uses squared deviations from the mean and squaring impacts small and large numbers differently. For example, squaring 3 produces 9, but squaring 6 produces 36. So a deviation that is twice as large produces a squared deviation that is four times as large. In a sense, standard deviation considers large deviations slightly more important than small deviations, whereas mean deviation treats them all the same.

The standard deviation you computed in Example 13 is also known as the *population standard deviation* and is often represented by σ (the Greek lowercase letter sigma). Here's the technical formula for population standard deviation. Note that, in this formula, the mean of the data set is the *population mean* and is represented by μ, the Greek lowercase letter mu (read "mew"):

$$\sigma = \sqrt{\frac{1}{N}\Sigma(\mu - x)^2}$$

This formula may look a bit intimidating, but it's just capturing the same painless steps outlined earlier in this section. First, find the mean of the data set, μ. Then, make a new list of the individual deviations from the mean, $(\mu - x)$. Next, square each of the numbers in that list, $(\mu - x)^2$. Then, compute their average, $\frac{1}{N}\Sigma(\mu - x)^2$. Finally, take the square root, $\sqrt{\frac{1}{N}\Sigma(\mu - x)^2}$.

In practice, the standard deviation is the preferred measure of deviation. Although the formula may seem more complicated than the formula for mean deviation, the standard deviation is more applicable in most mathematical and statistical contexts. The concept of mean absolute deviation is more straightforward than that of standard deviation—mean deviation is just the average distance to the mean for all the data points in the data set—but it is not as widely used.

Although the formula may look complicated, remember that standard deviation, like mean deviation, is simply a way to measure the average distance from the mean for all the data in a set. As such, it's an invaluable tool in understanding the spread of data, and along with the mean, it's a statistic you'll use in every chapter that follows. Although Example 13 showed you how to calculate the standard deviation by hand, usually you will use a calculator or computer program to compute the standard deviation of a data set.

It's important to understand how to use the formula for standard deviation, but most calculations in statistics are done with calculators or computers. Not only can calculators help with squaring and adding individual deviations, but many calculators have statistical capabilities that allow you to enter a list of numbers and compute standard deviation with a single command. Be sure to familiarize yourself with how your calculator works. It can make computing standard deviation painless!

There's also a second formula for standard deviation you'll encounter in statistics called the *sample standard deviation*. It's usually represented by an S (which stands for "sample"):

$$S = \sqrt{\frac{1}{n-1}\sum(\overline{x} - x)^2}$$

The formula for sample standard deviation looks very similar to the formula for population standard deviation. The only differences are dividing by $n - 1$ instead of n and writing $\overline{x}$ instead of μ. The formulas are different because the sample standard deviation has a different purpose than the population standard deviation. The population standard deviation, σ, is used to compute the standard deviaton of a set when you have all of the data. The sample standard deviation, S, is used to estimate the unknown population standard deviation from a particular sample. The context of the problem should make it clear which formula you want to use. You'll learn much more about this in Chapters 8 and 9, which cover sampling and inference. For now, though, here's a brief example of when to use the sample standard deviation rather than the population standard deviation.

Example 14:

Suppose you want to study the heights of adult American men. You want to know about the entire population of 100 million adult American men, but you can't possibly collect all that data. So instead you just study a sample of the population (maybe a data set with the heights of 75 adult American men).

Ideally, you would hope that the statistics of the sample—the mean, the deviation, and so on—would tell you about the statistics of the entire population. This is actually true of the mean. Under the right circumstances, the mean of the sample of 75 adult American men will be a good estimate of the mean of the entire population of 100 million adult American men. However, this is not true of the standard deviation. If you use the formula for σ, the population standard deviation, on the sample of 75 men, this will not be a good estimate of the standard deviation of the entire population. Instead, using S, the sample standard deviation, will provide a better estimate.

CAUTION—Major Mistake Territory!

You will usually use your calculator to compute standard deviation. Just be careful about using the correct one: σ for population standard deviation and S for sample standard deviation.

BRAIN TICKLERS Set #9

1. Consider the following two data sets:

   Data Set A: 10, 20, 30, 40, 50, 60, 70, 80, 90, 100

   Data Set B: 10, 10, 10, 10, 10, 100, 100, 100, 100, 100

Estimate which data set has the higher standard deviation. Then calculate the standard deviations of both sets, and check your estimate.

2. Below are two box plots that represent two data sets (A and B), each of which has 10 data points.

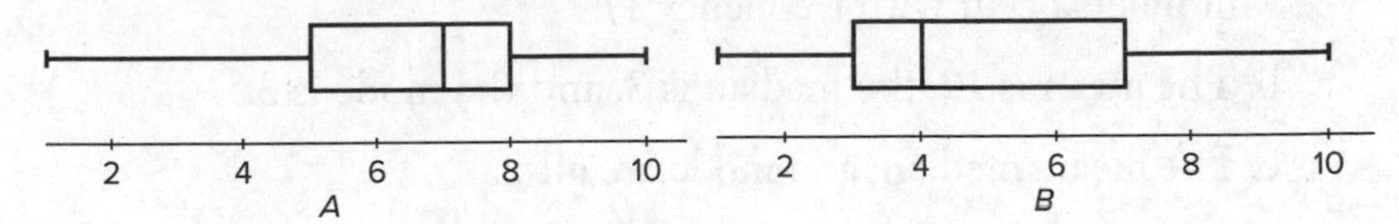

Which has the higher standard deviation?

3. The mean absolute deviation of the data set $\{a, b, c, d, e\}$ is 5.

   a. What is the mean absolute deviation of the data set $\{a + 3, b + 3, c + 3, d + 3, e + 3\}$?

   b. What is the mean absolute deviation of the data set $\{2a, 2b, 2c, 2d, 2e\}$?

4. Suppose you recorded the temperature inside your freezer once every hour for 24 hours and created a data set. In addition, you also recorded the temperature outside once per hour for 24 hours to create a second data set. Which data set do you think would have the higher standard deviation?

5. How would you expect the individual deviation of an outlier to compare to the standard deviation of a data set?

(Answers are on page 69.)

Brain Ticklers—The Answers

Set #7, pages 50–51

1. a. The mean is 550, the median is 550, and there is no mode (all numbers occur with frequency 1).

   b. The mean is 12, the median is 3, and the mode is 3.

   c. The mean, median, and mode are all 7.

2. There are many possible sets you could create. Two examples are $\{1, 2, 3, 21, 23\}$ and $\{1, 2, 3, 4, 40\}$.

3. a. 13. Shifting all the values by 3 shifts the mean by 3 as well.

   b. 9. Shifting all the values by 3 shifts the median by 3 as well.

   c. 20. Doubling all the values doubles the mean.

   d. 12. Doubling all the values doubles the median.

4. Since $a + b + c + d + e = 50$ and $a + b + c + d + e + f = 120$, then $f = 70$.

5. a. 3

   b. 50

   c. 13

Set #8, pages 59–60

1. a.

Min	Q1	Med	Q3	Max
–5	0	3	7	10

   b.

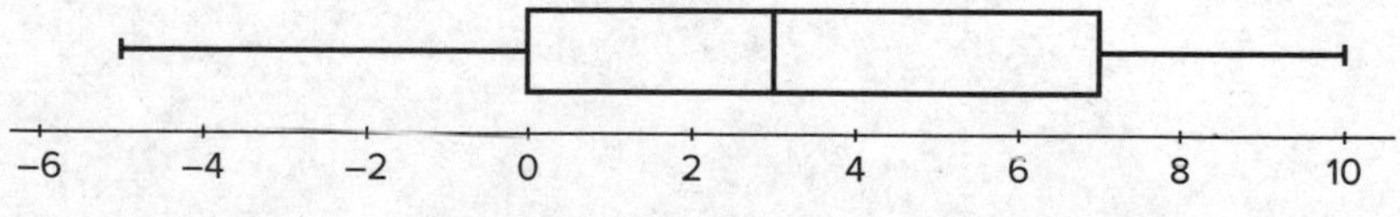

   –6 –4 –2 0 2 4 6 8 10

   c. The interquartile range is 7, so $1.5 \times IQR = 10.5$. This means that -5 is not an outlier because it is less than 10.5 away from the lower quartile of 0.

2. The upper quartile is equivalent to the 75th percentile, and the lower quartile is equivalent to the 25th percentile.

3. The range of Company X's stock price will probably be larger. If each company's stock changed by 1% over the course of the week, Company X's stock would change by $10, while Company Y's stock would change by 10 cents. In short, you might expect a higher-priced stock to fluctuate more than a lower-priced one, but of course, there will be exceptions.

Set #9, pages 66–67

1. It is reasonable to hypothesize that Data Set B has the higher standard deviation before any calculations have been performed. The standard deviation of Data Set A is 28.72. The standard deviation of Data Set B is 45. So the estimate was correct.

2. Box plot *B* has the higher standard deviation. The data is more spread out from the mean.

3. a. 5. Shifting every data point by 3 also shifts the mean by 3, so the deviations remain the same.

   b. 10. Doubling all the data values doubles the distances to the mean.

4. You would expect a higher standard deviation with the data taken from outside the house. Your freezer should stay at a consistent temperature, which means the data will not vary much. The temperature outside is much more likely to vary over the course of a full day. Since the standard deviation is a measure of how much the data varies about its mean, the standard deviation of the outdoor temperatures should be higher.

5. Since part of being an outlier is being far from the mean, the individual deviation of an outlier should be larger than the standard deviation of a data set, which is an average distance from the mean.

Chapter 4

Distributions of Data

To understand a data set, you need to understand its center, spread, and shape. In this chapter, you'll learn about the different shapes that data sets can take by exploring the most common distributions of data. Familiarity with these distributions makes understanding the shape of a data set easy and, together with summary statistics that describe center and spread, makes analyzing and interpreting data painless.

The Shape of Data

You can see the shape of a data set by graphing it. As you saw in Chapter 2, a frequency histogram is a graph that displays the values data takes and how often it takes them. A histogram paints a picture of how the data is distributed, that is, how it is spread out among its possible values. This picture of the data distribution helps you understand its important features.

Here is a histogram that shows data distributed along a horizontal axis from 0 to 100.

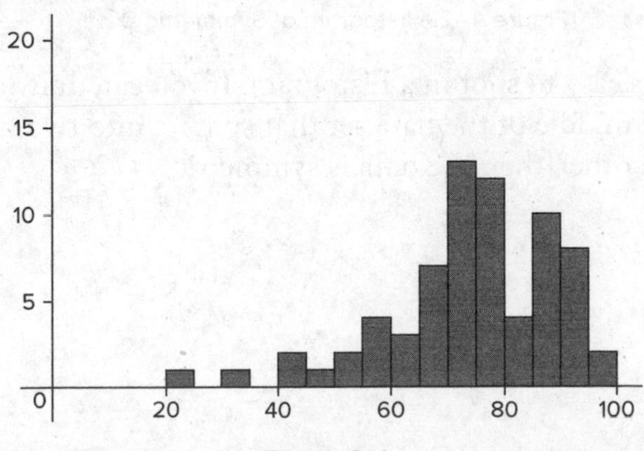

Figure 4–1

Even without any context, you can still identify features of the data just from the picture. You can see that most of the data is above 50, but there are some low data points in the 20s and 30s as well. Much of the data is clustered near 70, but there is another cluster near 90. With some experience working with data, you might even be able to estimate the mean (here, around 73) and median (here, around 76) of the data just from the shape of a histogram.

Knowing that this is a histogram of physics exam scores allows you to put the shape of the data in context. What does the shape of the data tell you about how students are doing in this class? About how effective the teacher is in teaching this class? About the quality of the exam? The shape of the distribution can lead you to pose useful questions about your data and identify areas for further analysis.

Symmetric Data Distributions

Here's an example of a histogram that displays one of the most useful features a data distribution can have: *symmetry*.

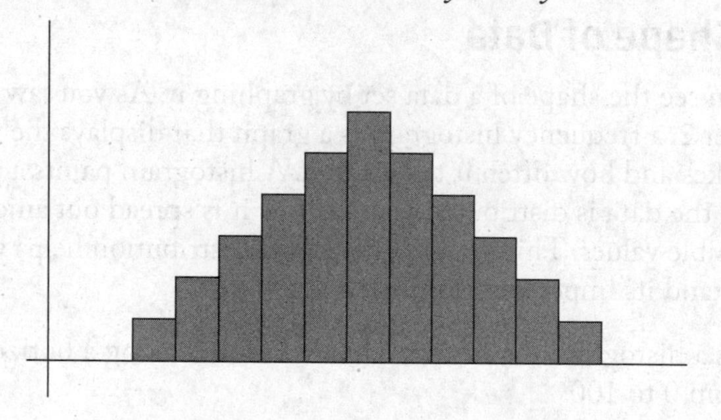

Figure 4–2. Histogram of Symmetric Data

Symmetry is easy to spot in a histogram. If you can draw a line through the middle of the data set that splits it into two mirror images of each other, then the data is symmetric.

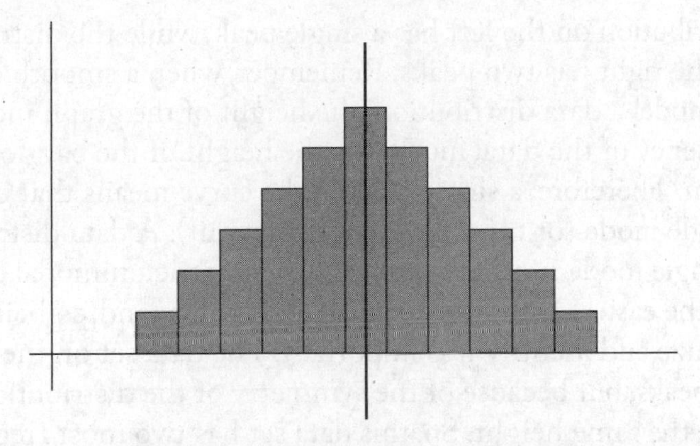

Figure 4–3. Histogram of Symmetric Data

Symmetric data can come in many forms. Here are some examples of symmetric data distributions, some of which are discrete and some of which are continuous.

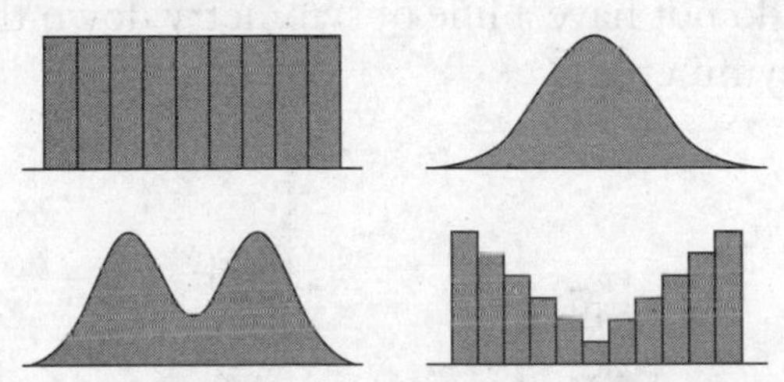

Figure 4–4. Examples of Symmetric Data

Data that is symmetric is easier to analyze and understand because it's made up of two identical parts. Whatever you learn about one half can be applied to the other, so in a sense, you only have to do half as much work to understand the whole data set. As you'll see later in this chapter, some of the most important data distributions in statistics are symmetric, and symmetry results in lots of useful properties.

Next, here are two continuous data distributions that are both symmetric but differ in an important way.

Figure 4–5. Unimodal and Bimodal Symmetric Data

The distribution on the left has a single peak, while the distribution on the right has two peaks. Remember, when a smooth curve is used to model a data distribution, the height of the graph indicates the frequency of the data, much like the height of the bar does in a histogram. Therefore, a single peak in the curve means that the data has a single mode (or most frequent data point). A data distribution with a single mode is called *unimodal*. Symmetric unimodal data is some of the easiest data to work with and understand, so being able to recognize and identify it is important. The data set on the right has two peaks, but because of the symmetry of the distribution, the peaks are the same height. So, this data set has two most frequent data points, or two modes, and for that reason it is called *bimodal*.

Skewed Data Distributions

Of course, not all data is symmetric. Here are some examples of distributions that do not have a line of symmetry down the middle and are therefore asymmetric.

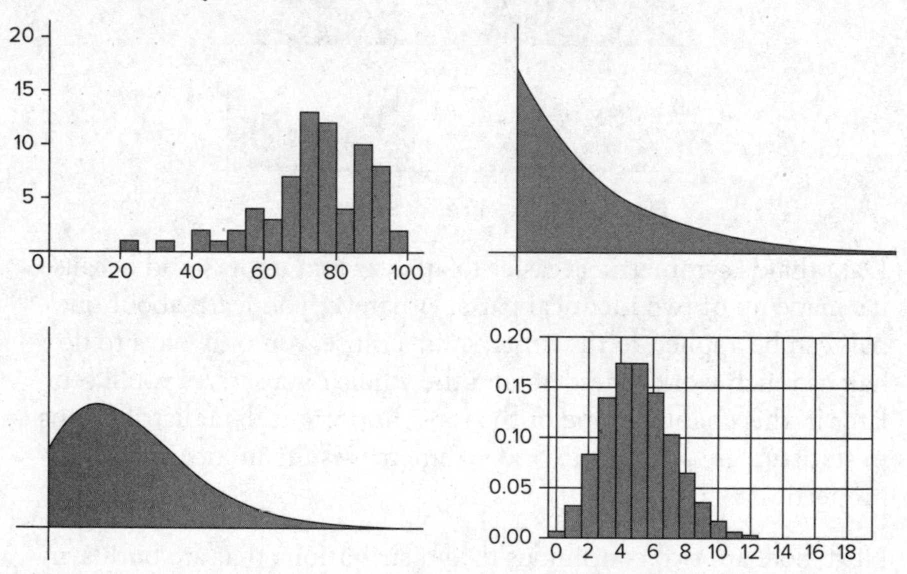

Figure 4–6. Examples of Asymmetric Data

These are examples of skewed data distributions. In a *skewed* data distribution, the data is stretched out more on one side than on the other.

You can also think of the asymmetry of skewed data as a difference in how the data "tails off" on either side. In the following distribution, the "tail" on the right side of the data set is longer than the tail on the left side.

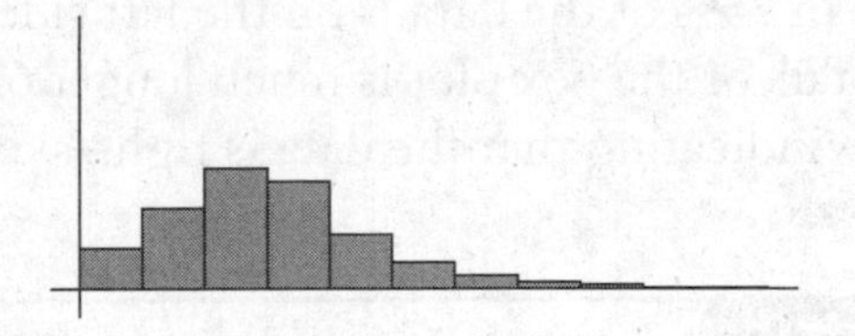

Figure 4–7. Histogram of Right-Skewed Data

When the tail on the right is longer than the tail on the left, you call the data *skewed right* or *right-skewed*. If the tail on the left is longer, then the data is *skewed left* or *left-skewed*, as shown below.

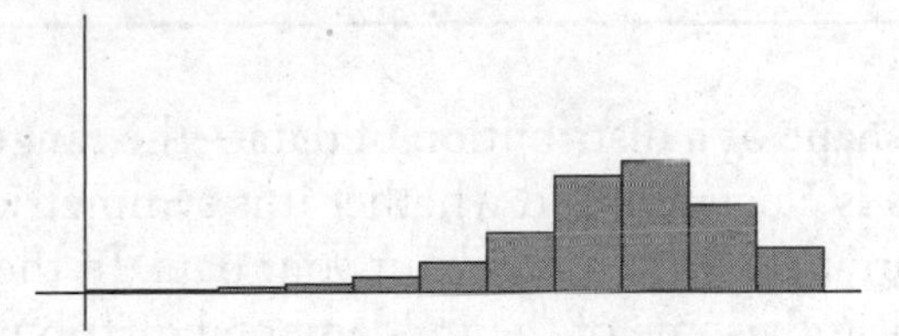

Figure 4–8. Histogram of Left-Skewed Data

CAUTION—Major Mistake Territory!

When a distribution is skewed left, most of the data appears on the right, and when a distribution is skewed right, most of the data appears on the left. This makes it easy to confuse the definitions of left-skewed and right-skewed data. To avoid this mistake, look for the tail of the data. The longer tail indicates the direction of the skew.

You can also recognize skewed data from its box plot. Here's the box plot for the right-skewed data set.

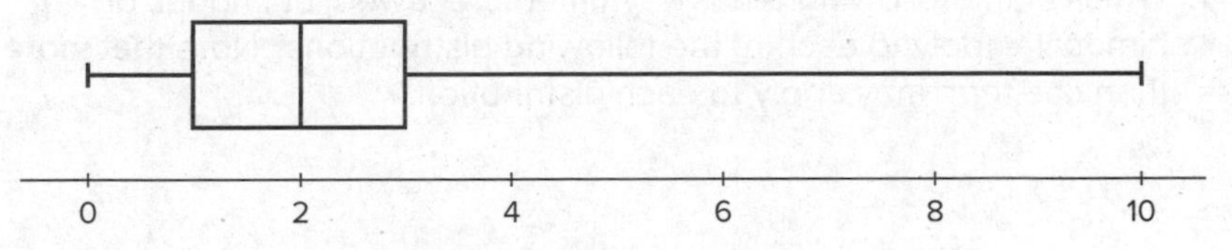

Figure 4–9. Box Plot of Right-Skewed Data

Recall that in a box plot, the box shows you where half the data lies. Here, the box is off center, which tells you that the data is skewed. (In fact, more than 75% of the data is on the left side of the graph.) The whisker, or tail, of the box plot is much longer on the right than on the left, again indicating that the data is right-skewed.

PAINLESS TIP

Starting with a symmetric distribution of data, if you grab the right end and stretch the data out (away from the center), you get *right-skewed* data. If you pull on the left end and stretch the data out, you'll get *left-skewed* data.

Knowing the shape of a distribution of data—the range it covers, where the data is clustered, and whether it is symmetric or skewed—provides important information about your data. In the next few sections, you'll study some of the fundamental distributions of data and learn how knowing about shape (together with the center and spread) can help you really begin to understand a data set.

CAUTION—Major Mistake Territory!

Many errors in statistics are made by assuming that data has features that it does not actually have, like symmetry. Be sure to investigate the shape of the data by graphing it.

BRAIN TICKLERS Set #10

1. Which of the following terms—symmetric, skewed, unimodal, or bimodal—apply to each of the following distributions? Note that more than one term may apply to each distribution.

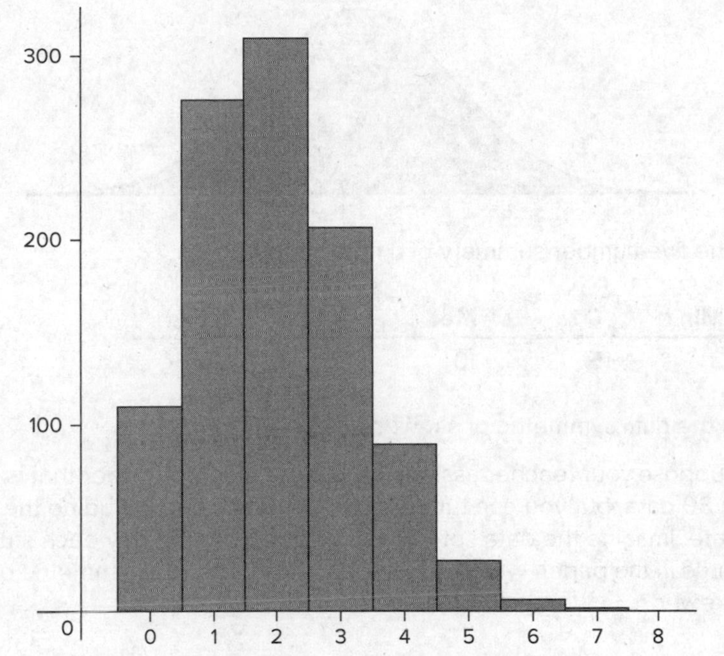

b.

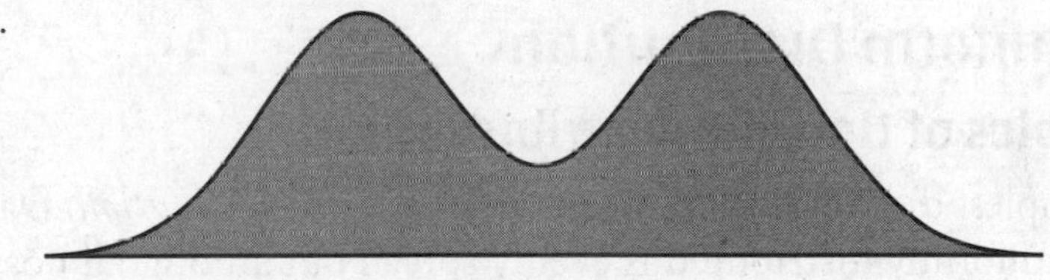

c.

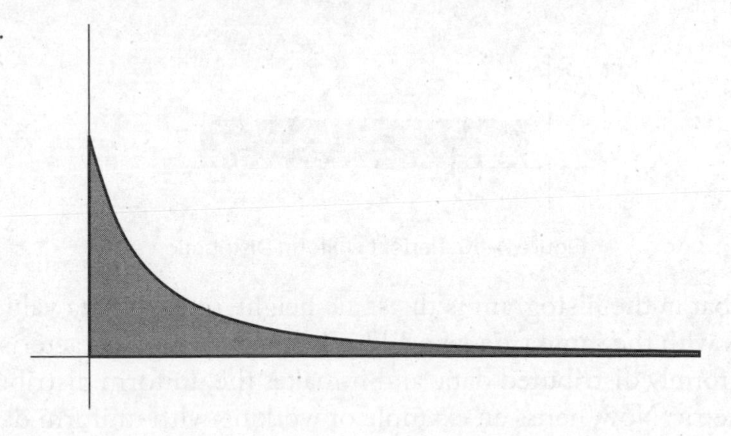

d.

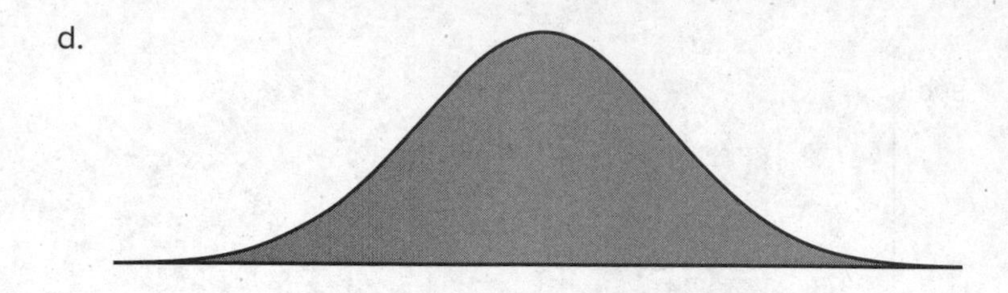

2. The five-number summary of a data set is shown.

Min	Q1	Med	Q3	Max
3	5	10	25	56

Is the data symmetric or skewed?

3. Suppose your teacher assigns a 20-page research paper that is due in 30 days, but you can turn it in any day up to and including the due date. Imagine the data set formed by recording the day each student turns in the paper. Would you expect this data to be symmetric or skewed?

(Answers are on page 100.)

The Uniform Distribution

Examples of Uniform Distributions

The simplest distribution of data is the *uniform distribution*. Data that is uniformly distributed is evenly spread out among all possible values. Here is a histogram of perfectly uniform discrete data.

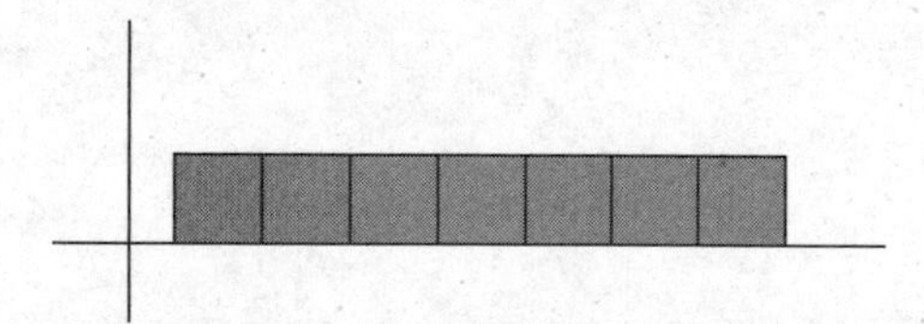

Figure 4–10. Perfect Uniform Distribution

Each bar in the histogram is the same height, so each data value occurs with the same frequency. This is the defining characteristic of uniformly distributed data, and it makes the uniform distribution symmetric. Now, here's an example of working with uniform data.

Example 1:

Recall the data set of the ages (in years) of 1,000 Americans that was presented in the following histogram in Chapter 2.

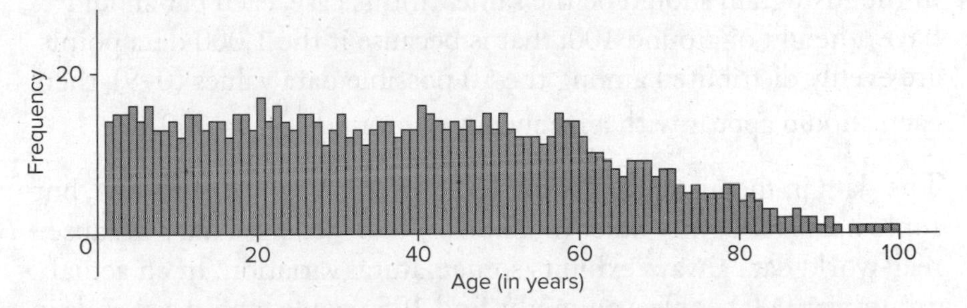

Figure 4–11. Age of 1,000 Americans

This data set is not uniform; you can tell this by the different heights of the bars. However, if you create a new data set by taking the ones digit from each age, you'll get a uniform data set. For example, if the age in the original data set was 35, put a 5 in the new data set. If the age was 21, put a 1 in the new data set. For an age of 80, put a 0 in your new data set, and so on. Here's what a histogram of this Ones Digits data would look like.

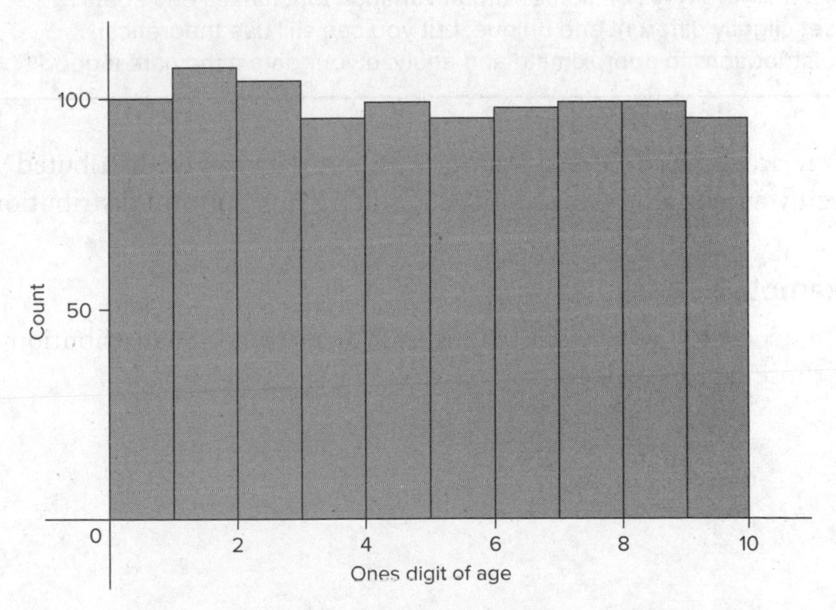

Figure 4–12. Age of 1,000 Americans

Notice that each number from 0 to 9 occurs with roughly the same frequency. That's because a person's age is just as likely to end in a 0 as it is to end in a 1 or a 7 or a 9. This means the height of each bar in the histogram should be the same. In this case, each bar should have a height of around 100; that is because if the 1,000 data points are evenly distributed among the 10 possible data values (0–9), then each should appear with a frequency of around $\frac{1{,}000}{10} = 100$.

The data in the histogram above looks pretty evenly distributed, but some bars are slightly taller (or shorter) than others. That's because real-world data always exhibits some natural variation. In an actual group of 1,000 people, you might find 105 people whose age ends in 0, 97 people whose age ends in 1, and so on. There will be slight differences, or variations, in the real-world data. Even so, you can still treat this data as uniformly distributed because it can be very closely approximated by this distribution.

PAINLESS TIP

Data from the real world rarely fits a theoretical distribution perfectly. There will always be some natural variation that makes each data set slightly different and unique, but you can still use theoretical distributions to approximate and analyze your data if they are a good fit.

Here are some additional examples of how data can be distributed evenly among all the possibilities, resulting in a uniform distribution.

Example 2:

You decide to flip a coin 1,000 times. What would the distribution of those flips look like?

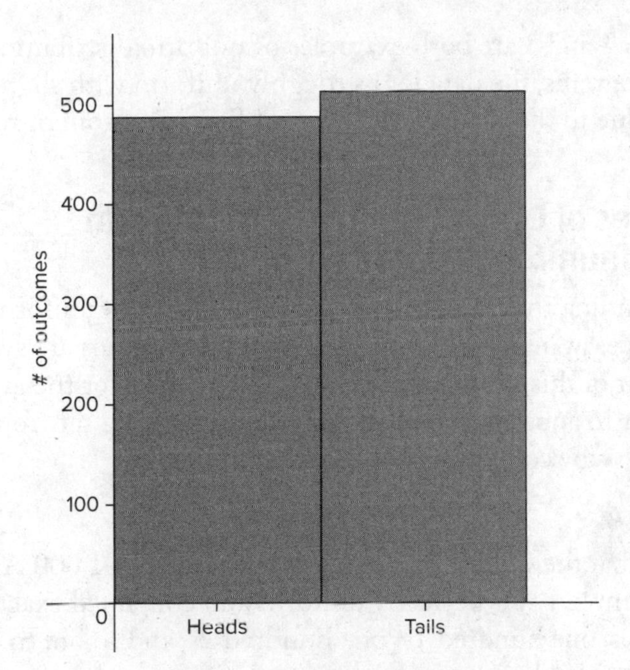

Figure 4–13. 1,000 Coin Flips

You would expect roughly 500 heads and 500 tails.

Example 3:

You decide to roll a fair six-sided die 1,000 times. What would the distribution of those rolls look like?

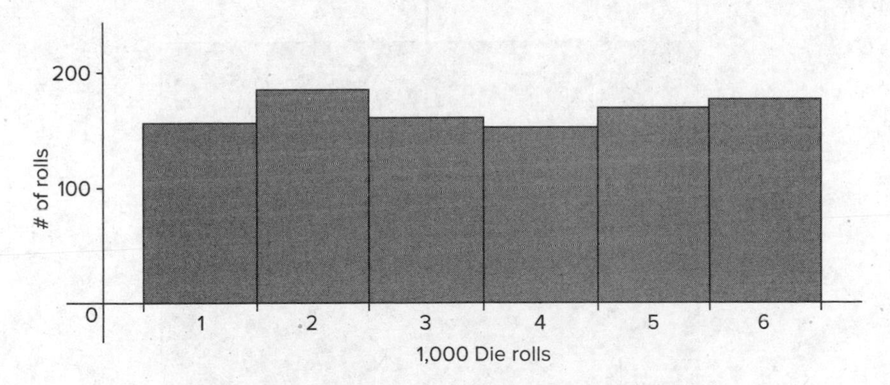

Figure 4–14. 1,000 Die Rolls

You would expect $\frac{1{,}000}{6} \approx 166.67$ or around 167 occurrences each of 1s, 2s, 3s, 4s, 5s, and 6s.

Examples 2 and 3 are both examples of uniform distributions. In both histograms, the data looks roughly uniform with slight imperfections due to the random variation of flipping a coin or rolling a die.

Measures of Central Tendency of Uniform (and Nonuniform) Distributions

Understanding the measures of central tendency of uniform distributions is easy, in part, because uniform distributions are symmetric. As a result of this symmetry, it would make sense for the average data point to appear right in the middle of the data set. You can confirm this by computing the mean.

Example 4:

Imagine that the Ones Digit data set of the ages of 1,000 Americans from Example 1 was perfectly uniform and contained exactly one hundred 0s, one hundred 1s, one hundred 2s, and so on. In that case, the mean would be:

$$\frac{(100 \times 0) + (100 \times 1) + (100 \times 2) + (100 \times 3) + (100 \times 4) + (100 \times 5) + (100 \times 6) + (100 \times 7) + (100 \times 8) + (100 \times 9)}{1{,}000}$$

This is equal to $\frac{4{,}500}{1{,}000} = 4.5$, making the average of the data set 4.5. Notice that 4.5 is right in the middle of the histogram.

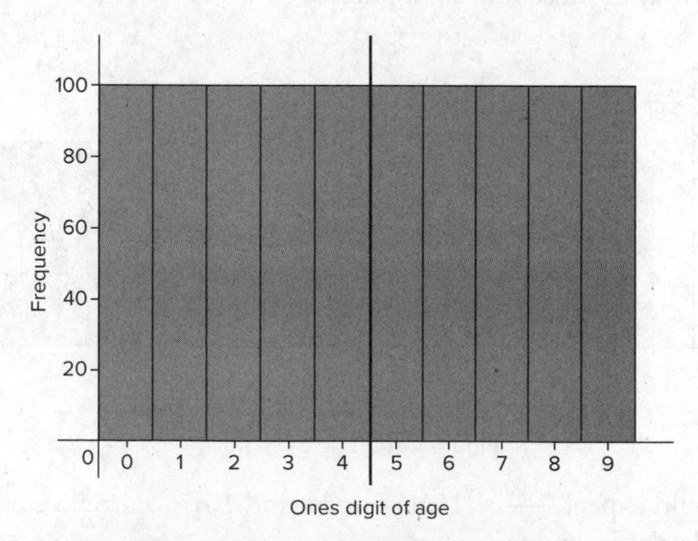

Figure 4–15. Age of 1,000 Americans

The average does not have to be a member of the data set. The same is true of the median.

In fact, since the data is uniform, the arithmetic can be made a little easier. Every digit occurs with the same frequency of 100, so you can factor out that 100 when computing the sum of all the numbers and then simplify the expression for the mean, as follows:

$$\frac{(100 \times 0) + (100 \times 1) + (100 \times 2) + (100 \times 3) + (100 \times 4) + (100 \times 5) + (100 \times 6) + (100 \times 7) + (100 \times 8) + (100 \times 9)}{1{,}000}$$

$$\frac{100(0+1+2+3+4+5+6+7+8+9)}{1{,}000}$$

$$\frac{0+1+2+3+4+5+6+7+8+9}{10} = \frac{45}{10} = 4.5$$

The mean of the Ones Digit data set is the same as the mean of the set of values (0, 1, 2, 3, . . . , 9) the data takes. This is because each value of the data (0, 1, 2, 3, . . . , 9) occurs with the same frequency in a uniform distribution.

The median is just as easy to understand in a uniform distribution. Since the data is evenly spread out, you would expect the middle number to be right in the middle of the distribution. Remember, to compute the median, first put the data in order and then find the middle value. Assuming the Ones Digit data is perfectly uniform, the list of 1,000 numbers in order would start with one hundred 0s, followed by one hundred 1s, then one hundred 2s, and so on, up to one hundred 9s. The 500th number of that list will be a 4, and the 501st number on the list will be a 5, so the median is 4.5, just like the mean.

When the data set contains an even amount of numbers, the median of the data set is the mean of the two numbers closest to the middle.

Notice that the median of the Ones Digit data set is just the median of the set

$$0, 1, 2, 3, 4, 5, 6, 7, 8, 9$$

which is 4.5. Just as with the mean, when finding the median of a uniform distribution, you can ignore the frequencies.

In summary, to compute the measures of central tendency for uniformly distributed data, just follow these painless steps.

Step 1: To find the mean, add up the possible values that the data can take and divide that sum by the number of possible values. (Ignore their frequencies.)

Step 2: To compute the median, arrange the possible values the data can take in order from least to greatest and find the middle value. If the set is even, take the average of the two values closest to the middle. (Again, ignore their frequencies.)

PAINLESS TIP

Since every data value appears with the same frequency in a uniform distribution, there is no mode.

Since the median is the same as the mean, the line drawn down the middle of the uniform distribution passes through both. This will always happen with perfectly uniform distributions, and it is a consequence of the symmetry of the distribution. Every number on one side of the line of symmetry has a mirror image partner on the other side. Because of the way mean and median are calculated, this property forces both of these measures of central tendency to end up on the line of symmetry.

When data is not uniform, the measures of central tendency probably won't be the same, as you'll see in the following example.

Example 5:

Imagine that you once again have recorded the ages of 1,000 Americans, but this time, instead of taking the ones digit from each age, you took the tens digit. For example, if the original age was 37, you put a

3 in a new data set. If the original age was 56, you put a 5 in that data set, and so on. (For an age under 10, the tens digit will be 0.) Call this the Tens Digit data set. Here's the histogram of that data.

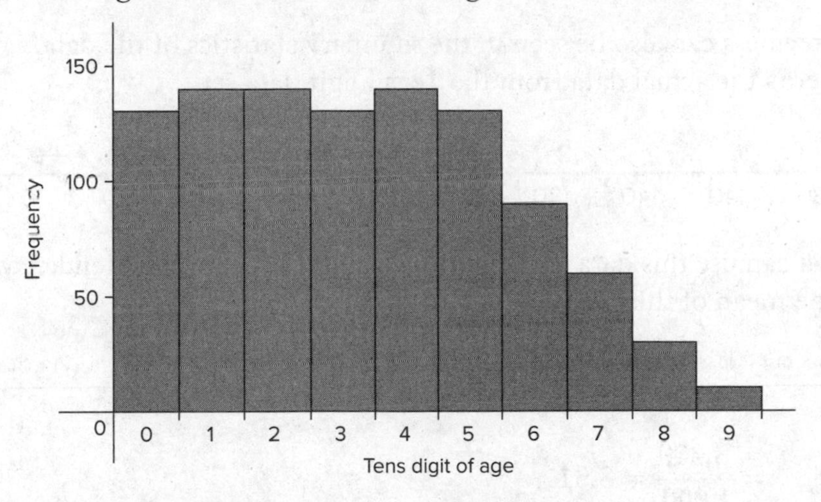

Figure 4–16. Age of 1,000 Americans

Notice that, unlike the Ones Digit data set, this Tens Digit data is not uniformly distributed. The heights of the bars are roughly even from 0 up to 5, but after that, the bars start to decrease in height. This data set is skewed right: the tail of the data stretches out on the right side.

It makes sense that this data is skewed. Living to 80 years old is less common than living to 20 years old. So in a random group of 1,000 people, there shouldn't be as many people in their 80s as there are people in their 20s. This means that far fewer people will have an 8 as the tens digit of their age than those who have a 2 as the tens digit of their age.

A data set that is skewed isn't distributed in the same way on either side of a center line. The bulk of the data is on one side or the other. You can see that in this histogram, but you can also see that in the box plot of this data.

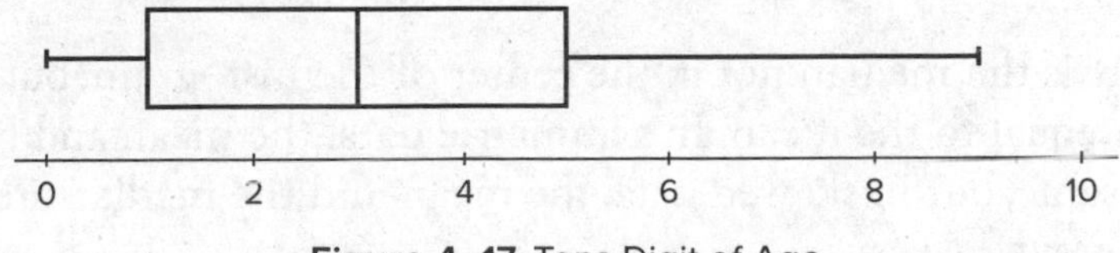

Figure 4–17. Tens Digit of Age

Notice how the box, which contains 50% of the data, lies toward the left end of the range and the longer tail stretches out to the right, making this right-skewed data.

Skewness can also be seen in the summary statistics of the data. Here's the actual data from the Tens Digit data set.

0	1	2	3	4	5	6	7	8	9
130	140	140	130	140	130	90	60	30	10

You can use this data to compute the measures of central tendency. The mean of this data set is:

$$\frac{(130 \times 0) + (140 \times 1) + (140 \times 2) + (130 \times 3) + (140 \times 4) + (130 \times 5) + (90 \times 6) + (60 \times 7) + (30 \times 8) + (10 \times 9)}{1{,}000}$$

$$= \frac{3{,}310}{1{,}000} = 3.31$$

The mean of the data, 3.31, is off center. The median of this data set, 3, is also off center.

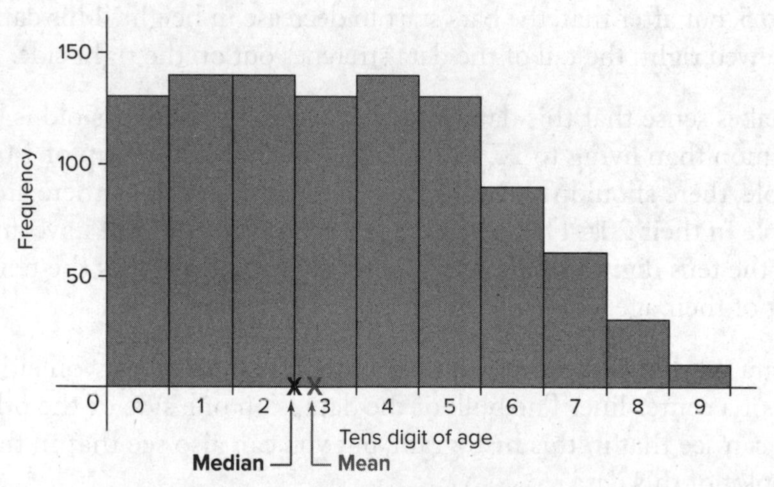

Figure 4–18. Age of 1,000 Americans

Not only is the median not in the center of the histogram, but it's also not equal to the mean. In symmetric data, the mean and median are the same, but in skewed data, the mean and the median are usually different.

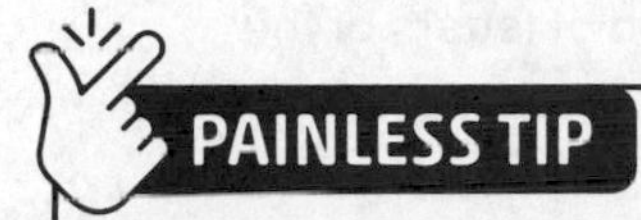

In right-skewed data, the mean is usually to the right of the median, and in left-skewed data, the mean is usually to the left of the median. A simple rule of thumb is that the mean usually follows the tail of the data.

When the mean and median of a data set are different, that tells you something important about your data. If the median is less than the mean, then you know that more than half the data lies below the mean. You saw an example of this in Chapter 3, Example 4, which discussed home prices. In the neighborhood with 90 homes that cost $100,000 each and 10 homes that cost $2,000,000 each, almost all the home prices were below the mean price of $290,000. In that data, the handful of high-priced homes pulled the mean (which is nonresistant to extreme values) above the median (which is resistant to extreme values), indicating that the data may be right-skewed.

In some situations, this difference between the mean and median can lead to what seems like a paradox. For example, in this same neighborhood, one might say that almost all the homes are "below average." Saying that almost everything is "below average" might sound strange at first, but when you translate this into the statement "almost all the data is below the mean," it sounds less paradoxical.

When the mean and the median of a data set are equal, however, that often indicates that the data is symmetric, like in the uniform distribution. In the next section, you'll see just how useful this kind of symmetry can be in a data distribution.

BRAIN TICKLERS Set #11

1. 1,000 data points are perfectly uniformly distributed from the numbers 1 to 20.

   a. What is the frequency of 7 in the data set?

   b. What is the mean of the data set?

   c. What is the five-number summary for the data set?

2. Data Set A is a uniform distribution from 0 to 10. Data Set B is a uniform distribution from 0 to 20. Is the standard deviation of Data Set B less than, equal to, or greater than the standard deviation of Data Set A?

3. In a book written in English, will the distribution of letters of the alphabet used be uniform or not uniform?

(Answers are on page 100.)

The Binomial Distribution

An Example of a Binomial Distribution

The *binomial distribution* is a discrete distribution that is unimodal; that is, all three measures of central tendency of the data set (the mean, median, and mode) occur around the single peak. Under certain circumstances, the binomial distribution is symmetric. Even when it isn't symmetric, it is often symmetric enough to be treated as though it were. These features make the binomial distribution easy to work with, which is especially important because of the fundamental role it plays in a variety of situations throughout math and statistics.

A simple way to begin to understand the binomial distribution is with an example involving coin flips.

Example 6:

Imagine repeatedly flipping a fair coin 10 times and counting the number of times the coin comes up tails. You might get 3 tails out of 10 flips the first time, 6 tails out of 10 flips the second time, 8 tails the third time, and so on. If you did this 1,000 times and graphed the data in a histogram, you'd see something like this.

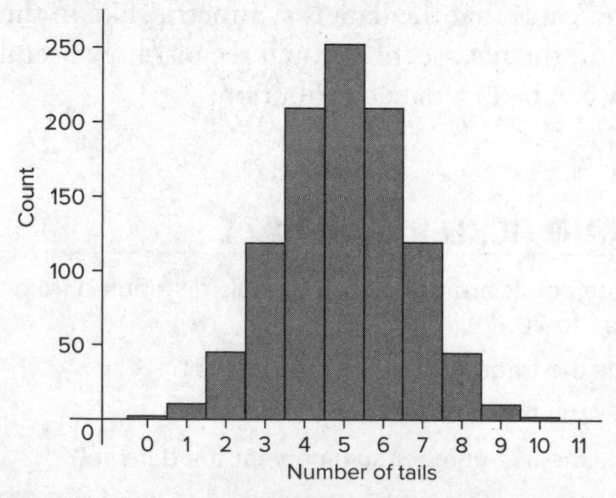

Figure 4–19. 1,000 Repetitions of 10 Flips of a Fair Coin

This is an example of binomially distributed data. The information you are collecting in each set of 10 coins flips is the count of the number of tails seen, so this is count data. Count data is frequently binomially distributed. Since counting things is a very common technique in statistics, this is one of the reasons why the binomial distribution comes up so often.

You can see that this data distribution is unimodal and symmetric. The line of symmetry runs through the single peak at 5, and so the mean, median, and mode of this binomially distributed data are all 5 tails. Since a fair coin should come up heads or tails with equal likelihood, it makes sense that 5 is the center of the data set. In 10 flips, you would expect to get tails half the time, which would be 5 tails. Of course, if you were to do this repeatedly, you would sometimes see more than 5 tails and sometimes see less than 5 tails. This is due to *random variation*. The outcome of the coin toss is a random process, so the outcomes of the individual flips of the coin can't be predicted. However, over time, statistical patterns will emerge. These patterns that arise from random variation are captured by the histogram in the bars of different heights, and the "average" outcome of 5 tails is also represented in the histogram as the single central peak.

The fact that a fair coin toss should come up heads half the time and tails half the time accounts for the symmetry of this distribution. The data ranges from 0 to 10 because the minimum number of tails is 0 and the maximum number is 10. Notice how the bar above 0 is about the same height as the bar above 10. This is because a count of 0 tails should occur about as often as a count of 10 tails. The reason for this is that when flipping a fair coin, 0 tails should happen about as often as flipping 0 heads, and flipping 0 heads is the same outcome as flipping 10 tails. So, flipping 0 tails and flipping 10 tails should occur with the same frequency. The same argument applies to 1 tail and 9 tails, 2 tails and 8 tails, and so on. This explains why this distribution is symmetric. Each number of tails in the distribution has a mirror image partner, and the mirror is located at 5, which is the mean, median, and mode of the data set.

If the coin you were flipping was unfair, then the distribution wouldn't be so symmetric. For example, suppose the coin was weighted so that it came up tails 80% of the time and heads 20% of the time. Doing the same thing as before (1,000 repetitions of flipping a coin 10 times and counting the number of tails) would produce a data set whose histogram looks like this (which you can see is not symmetric).

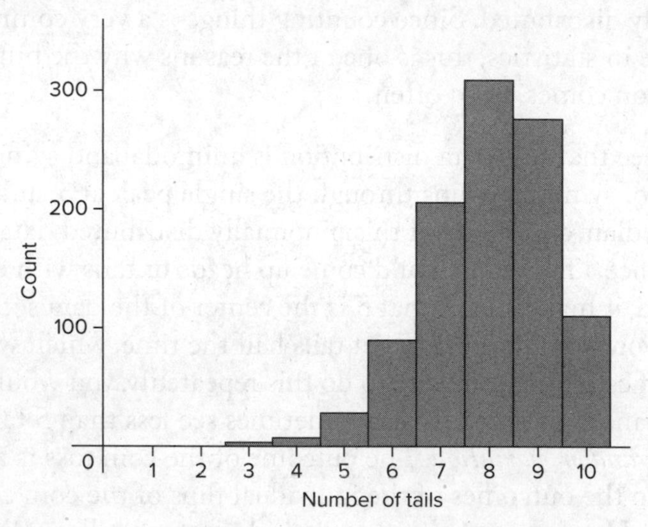

Figure 4–20. 1,000 Repetitions of 10 Flips of an Unfair Coin

When the coin is fair (and heads and tails come up with the same frequency), the distribution is symmetric, but when the coin is unfair, the distribution is skewed. Since this coin is much more likely to come up tails than heads, you would expect to count more tails than heads in 10 coins flips. This suggests that the mean count of tails should be higher than 5.

In fact, the mean count of tails in this distribution is 8, which makes remarkable sense. This unfair coin comes up tails 80% of the time, and since 80% of 10 is 8, you would expect 8 out of every 10 flips to be tails. Again, sometimes you will count more and sometimes less due to random variation, but on average, you'll count 8 tails.

This method works in general for computing the mean of a binomial distribution. Given an experiment that is successful p% of the time, if you perform that experiment n times, the mean number of successes you will see is p% of n. In the first part of the example, the experiment was flipping a fair coin and a success was flipping tails. Since flipping tails had a success rate of 50%, the mean number of successes you would see in 10 flips was 50% of 10, or 5. In the second part of the example, the experiment was flipping an unfair coin and the success of flipping tails was 80%. So the mean number of successes in 10 flips was 80% of 10, which is 8.

Measures of Central Tendency and Standard Deviation for Binomial Distributions

All of the preceding information can be summarized in a simple formula for the mean of a binomial distribution:

$$\mu = np$$

Here, μ is the mean, n stands for the number of *trials* (i.e., the number of times the experiment is repeated), and p stands for the *probability of success* of each trial. You'll learn more about probability in Chapters 6 and 7, but for now you can just think of it as the percent chance that what you're looking for occurs. Also, for this formula to apply, each trial has to be *independent* of each other, which just means that the outcome of one flip doesn't impact the outcome of another flip. You'll learn more about independence in Chapter 7.

This simple way to calculate the mean is one of the remarkable properties of count data and the binomial distribution, but there's more. There is also a simple formula for the standard deviation of a binomial distribution. The n and p here are the same as in the formula for the mean:

$$\sigma = \sqrt{np(1-p)}$$

With this formula, you can easily calculate the standard deviations for both of the earlier situations (flipping a fair coin and flipping an unfair coin). When the fair coin is flipped 10 times ($n = 10$) at a success rate of 50% ($p = 0.50$) each time, the standard deviation is around $\sigma = \sqrt{10 \times (0.5)(1-0.5)} \approx 1.58$ tails. When the unfair coin is flipped 10 times ($n = 10$) at a success rate of 80% ($p = 0.80$) each time, the standard deviation is around $\sigma = \sqrt{10 \times (0.8)(1-0.8)} \approx 1.26$ tails. Remember that the standard deviation is a measure of the dispersion of the data. If you look at the histograms of the two data sets, you'll see that with the unfair coin, the data is more tightly clustered around its mean. This is because the standard deviation is lower.

CAUTION—Major Mistake Territory!

Be aware that every distribution of data has its own formulas for mean and standard deviation. Don't make the mistake of applying the formula for the mean of a binomial distribution to some other distribution of data.

Yet another remarkable fact about the binomial distribution is that the mean, median, and mode are always roughly the same, even when the distribution is skewed. When the probability of success is 50%, the distribution is symmetric, so it seems natural that all three measures of center are the same. However, this is true of the binomial distribution even when the data is not symmetric. Even though the binomial distribution can be skewed when the probability of success isn't 50%, as long as the number of trials (n) is large enough, the distribution can be closely approximated by a symmetric distribution. For example, an unfair coin that comes up tails 80% of the time produces a skewed distribution, but if enough trials are conducted, the data can effectively be treated as though it is symmetric. This means that in many circumstances in the application of statistics, you can work with binomially distributed data as though it's approximately symmetric, even if it isn't perfectly so.

Of course, the binomial distribution isn't just about coin flips. It can be applied anytime you are counting the results of an action whose results can be classified as a success or failure (like making a sale in business, testing a drug in medicine, or checking for defects in a manufacturing process). The binomial distribution plays a fundamental role in sampling and inference, which you'll learn all about in the second half of this book.

BRAIN TICKLERS Set #12

1. Suppose you flip 15 fair coins and record the number of tails you see. Now, imagine repeating this a total of 1,000 times. What would you expect the average number of tails to be? Assume the data is binomially distributed.

2. Suppose you flip 15 fair coins and record the number of tails you see. Now, imagine repeating this a total of 1,000 times. What outcome would occur with roughly the same frequency as the outcome of 4 tails? Assume the data is binomially distributed.
3. Suppose you are in charge of quality control at a computer chip manufacturer, and an experimental new method produces defective chips at a rate of 5%. Suppose every day you select 50 chips at random and test them for defects. Assume this data is binomially distributed. How many defective chips would you expect to find each day, on average?
4. Use the formula $\sigma = \sqrt{np(1-p)}$ and a calculator to compute the population standard deviation of a binomially distributed data set for the following values:

   a. $n = 100; p = \frac{1}{2}$

   b. $n = 100; p = \frac{1}{10}$

   c. $n = 100; p = \frac{1}{100}$

   d. What appears to happen to the standard deviation as p gets closer to 0? What about when p gets closer and closer to 1?

(Answers are on page 100.)

The Geometric Distribution

An Example of a Geometric Distribution

The *geometric distribution* is another important data distribution. As with the binomial distribution, a good place to begin exploring the geometric distribution is through coin flips, but a different scenario is required to produce geometrically distributed data.

Example 7:

In the experiment of flipping a fair coin, call flipping tails a "success." To generate a new data set, you will flip a coin repeatedly until you see a successful outcome, that is, until you flip tails. Once you see tails, you will record the number of failures (the number of times you flipped heads before finally seeing tails) in a data set. For example, if you flip tails immediately, you record a 0 in the data set because you experienced zero failures—that is, zero heads—before seeing tails. If the sequence of coin flips proceeds heads, heads, heads, tails, you record a 3 in the data set (for the three heads you saw before you saw tails).

Imagine doing this many, many times and recording the results. This data would be geometrically distributed, and a relative frequency histogram would look like this.

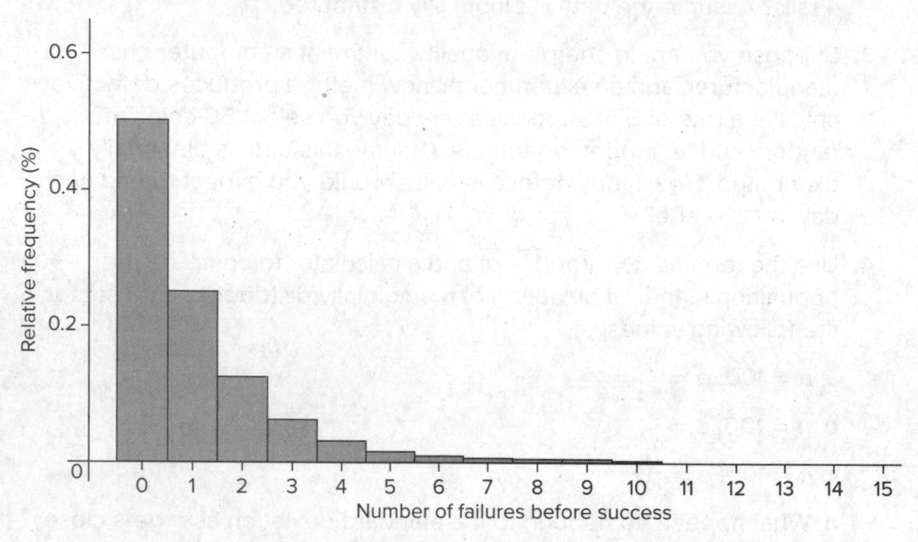

Figure 4–21. A Geometric Distribution

In a relative frequency histogram, the height of each bar represents the percentage of data for each corresponding value, so the vertical axis goes from 0 to 1 (where 1 is 100%).

The geometric distribution is a skewed distribution. You can see that the tail of the distribution extends to the right, making this right-skewed data. The skew is a consequence of how the data is constructed. With a fair coin, about half the time, you will see tails on the first toss, so the frequency of 0's in the histogram will be around 50%. Now, if your first flip is heads, then you flip again, about half of those second flips will be tails. In this case, you put a 1 in your data set. So 1 will have a frequency of around half of 50%, or 25% of the total tosses. The pattern in this data is that each bar in the histogram will be about half as high as the previous bar. The farther right you go, the shorter the bars will be, until they appear to vanish. There does not appear to be any bar above 10, which makes sense because even though it is theoretically possible to flip heads 10 times in a row before flipping tails, it is very, very unlikely.

A box plot can also show you the skew in a set of geometrically distributed data. Here's a box plot of 1,000 repetitions of the experiment of flipping a fair coin until tails appears.

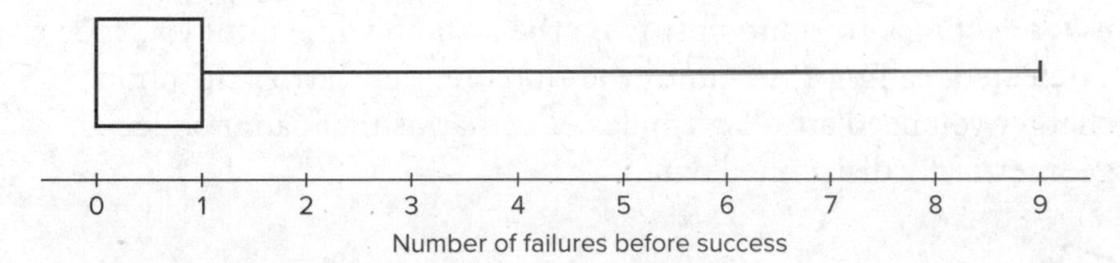

Figure 4–22. Box Plot of Geometric Data Representing 1,000 Repetitions

Notice how the box, which represents 50% of the data, is all the way to the left and the tail extends to the right in this right-skewed data.

Measures of Central Tendency and Standard Deviation for Geometric Distributions

Just like there was for the binomial distribution, there are also nice formulas for the important summary statistics of the geometric distribution. The mean of a geometric distribution is given by:

$$\mu = \frac{1}{p}$$

The standard deviation of a geometric distribution is given by:

$$\sigma = \frac{\sqrt{1-p}}{p}$$

The p is the same as in the formulas for the binomial distribution: it is the probability of success of each independent experiment. In the case of the fair coin flip, $p = \frac{1}{2}$, so the mean number of heads you'll encounter before you flip tails is around $\frac{1}{\frac{1}{2}} = 2$. The standard deviation of this data is $\sigma = \frac{\sqrt{1-0.5}}{0.5} = 1.41$ heads. These formulas make computing the descriptive statistics of the geometric distribution painless, but as always, remember that real-world data will display some random variation.

As with the binomial distribution, the geometric distribution applies to a lot more than coin flips. Any time you are repeating some independent action until a desired outcome occurs, you are generating geometrically distributed data. Rolling a pair of dice until the sum is 7, selecting a toy randomly from the assembly line until you spot a defect, or calling stores until you find one that carries the phone charger you need are all examples of scenarios that can produce geometrically distributed data.

BRAIN TICKLERS Set #13

1. Suppose you roll a fair six-sided die until you see a 6 appear. On average, how many non-6s will you see before your first 6 appears?
2. Suppose a fair coin is flipped until you see tails, and then the total number of flips is recorded. Will the median of this data set be closer to 1 or to 5?
3. In a geometric distribution, would you expect the mean to be less than, equal to, or greater than the median?

(Answers are on page 101.)

Other Distributions

Here are some other distributions that you may encounter as you continue your study of statistics.

The Exponential Distribution

In the previous section, you learned about the geometric distribution, which is a discrete data distribution. The related continuous distribution is called the *exponential distribution*, and its graph looks like this:

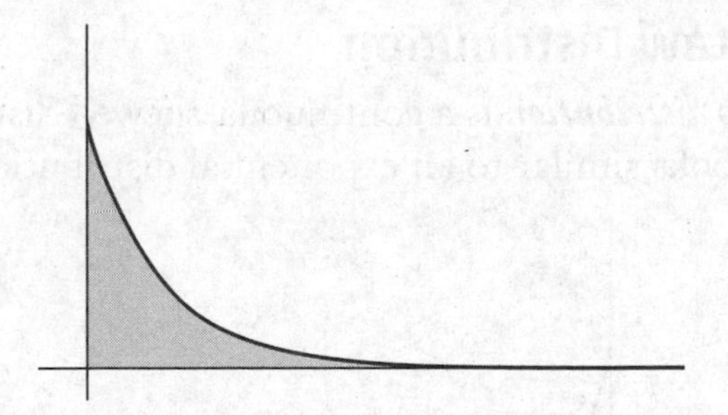

Figure 4–23. Example of an Exponential Distribution

Exponential distributions are used to model many real-world phenomena, like how long you have to wait for a bus on the street, a hospital bed in an emergency room, or a customer service representative on the phone. A continuous distribution is useful in these cases because continuous data can take values in between whole numbers. You might wait for a bus for 1.3 minutes, or 1.76 minutes, or 12.453 minutes, and so on.

Like the geometric distribution, the exponential distribution is skewed and has a tail that extends far to the right. The exponential distribution has a long tail for much the same reason the geometric distribution has one. Just as it's possible, though very unlikely, to flip heads 10 times before first seeing tails, it's possible, but very unlikely, that you would have to wait 45 minutes for a bus to arrive (although you'd probably start walking before that). In fact, the frequency of the exponential distribution decays at an exponential rate, which is where the name of the distribution comes from. This is similar to how the bars of the histogram of the geometric distribution shrink by a common factor.

1+2=3 MATH TALK!

Exponential growth (or decay) means that the quantity is multiplied (or divided) by the same amount again and again. For example, the pattern 1, 2, 4, 8, 16, 32, . . . is an example of exponential growth: you multiply each number by 2 to get the next number. The pattern 10,000; 1,000; 100; 10; 1; 0.1; . . . is an example of exponential decay: you divide by 10 each time to get the next number in the sequence.

The Power Law Distribution

The *power law distribution* is a continuous, skewed distribution whose graph looks similar to an exponential distribution.

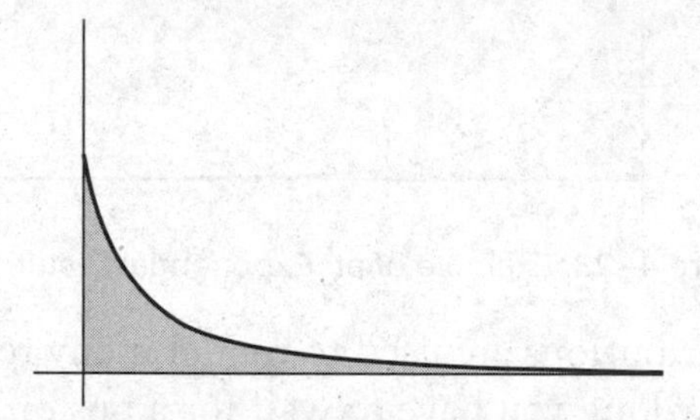

Figure 4–24. Example of a Power Law Distribution

Though similar in appearance, power law distributions differ from exponential distributions in how rapidly the frequencies decline along the horizontal axis. Frequencies drop to 0 very rapidly in an exponential distribution (as in the geometric distribution from Example 7, in which each bar was half the height of the previous bar), but they decline much more slowly in a power law distribution.

Like exponential distributions, power law distributions frequently pop up in nature. You might see a power law distribution when looking at how many connections people have on social networks, how food sources are interconnected in an ecosystem, or how frequently words are used in a language. The power law distribution is an advanced topic, but it arises so frequently in the application of statistics to the sciences that it is worth being familiar with.

The Poisson Distribution

The *Poisson distribution* is a discrete distribution that is used to model how frequently certain events occur in a fixed time frame. A histogram of a Poisson distribution might look like this:

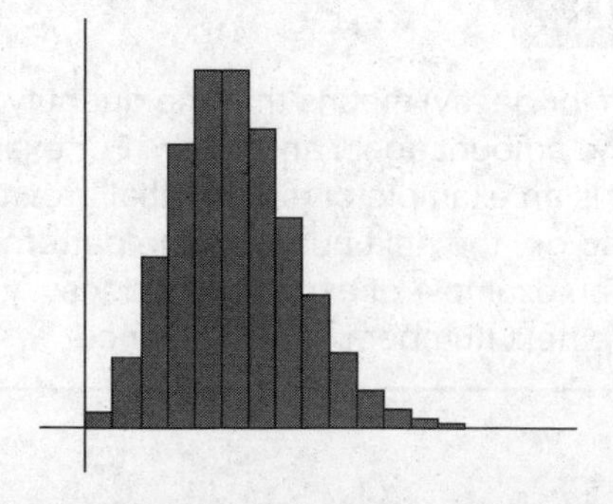

Figure 4–25. Example of a Poisson Distribution

You can see that the Poisson distribution is a skewed distribution, but how skewed it is depends on the nature of the phenomenon being modeled. The number of customer service calls a company receives each hour, the number of meteorites that strike Earth each year, or the number of Category 5 hurricanes a particular region experiences every century all might be modeled with the Poisson distribution. The Poisson distribution is another advanced topic in statistics, but it, too, is worth knowing.

The Normal Distribution

The *normal distribution* is a continuous, symmetric, unimodal distribution of data whose graph has the classic "bell-shaped" curve.

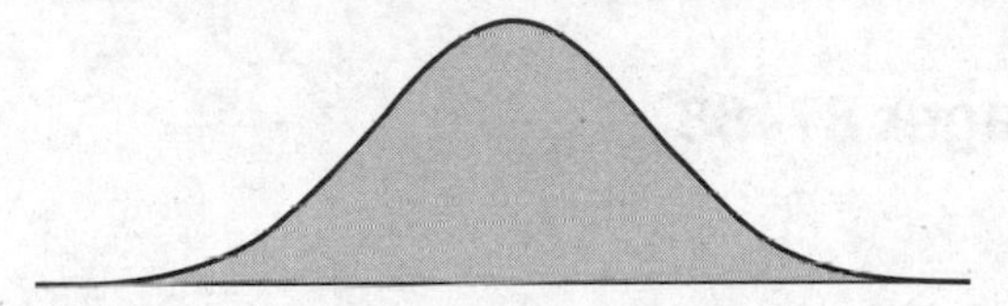

Figure 4–26. Example of a Normal Distribution

This distribution is used to model many different natural phenomena. It is so important and ubiquitous in statistics that it is the focus of the next chapter.

Brain Ticklers—The Answers

Set #10, pages 76–78

1. a. Skewed, unimodal

   b. Symmetric, bimodal

   c. Skewed, unimodal

   d. Symmetric, unimodal
2. Skewed (right-skewed, in fact)
3. Skewed. Some students will turn in their papers early, but most students will probably turn in their papers on the final day.

Set #11, pages 87–88

1. a. 50

   b. 10.5

   c. 1, 5.5, 10.5, 15.5, 20
2. The standard deviation of Data Set B is greater than that of Data Set A. In Data Set A, the maximum deviation is five. Much of the data in Data Set B has a deviation greater than 5, so the average deviation will be above 5.
3. Not uniform. For example, the letters *r*, *s*, and *t* will likely appear much more frequently than the letters *q* and *z*.

Set #12, pages 92–93

1. 7.5
2. 11
3. 2.5
4. a. 5

   b. 3

   c. 0.99

   d. As p gets closer to 0, the standard deviation gets smaller. The standard deviation also gets smaller as p gets closer to 1.

Set #13, page 96

1. 6. Since $p = \frac{1}{6}$.
2. Closer to 1. About half of the first flips should be tails, so about 50% of the data will be 1.
3. The mean should be greater than the median in a geometric distribution. The longer tail to the right will likely pull the mean above the median.

Chapter 5

The Normal Distribution

Statistics is an applied science, so once you understand the basics of statistics, you'll want to apply what you've learned. You'll look around in the world and see data—times, prices, populations, ages, revenues—and you'll use your knowledge of statistics to make sense of it.

Data in the real world comes in many shapes and many distributions; you were introduced to many different distributions in Chapter 4. However, when it comes to data in the real world, one of the most common, and useful, shapes is the normal distribution.

Normally Distributed Data

The Shape of the Normal Distribution

The *normal distribution* is a continuous data distribution that looks like this:

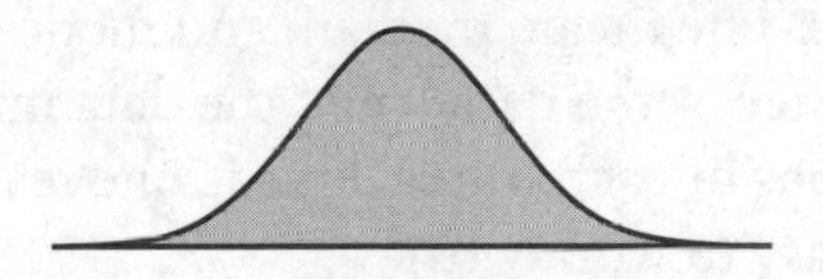

Figure 5–1. The Normal Distribution

As you can see from the above graph, the normal distribution is symmetric and unimodal. It's symmetric because you can draw a line down the middle and see the same shape of data on either side. It's unimodal because the graph of the data has only one peak, which means there is only one mode.

> The normal distribution is also known as the *Gaussian distribution*, after the mathematician Carl Friedrich Gauss.

Many kinds of real-world data, from characteristics like height to performance indicators like test scores, are distributed in a way that is approximately normal. Here's what a histogram of data that is approximately normal might look like:

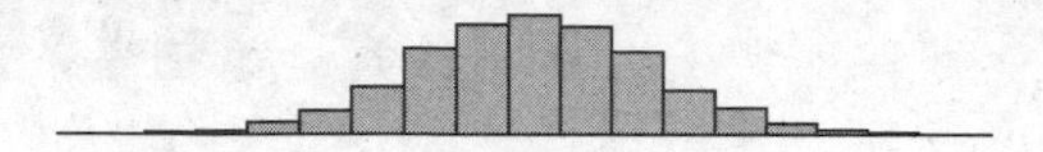

Figure 5–2. Histogram of Data that is Approximately Normal

You can see that this histogram shares a similar shape to the graph of the normal distribution. It's possible to find a continuous normal curve that closely fits this data, which means you can approximate this discrete data with a continuous normal distribution.

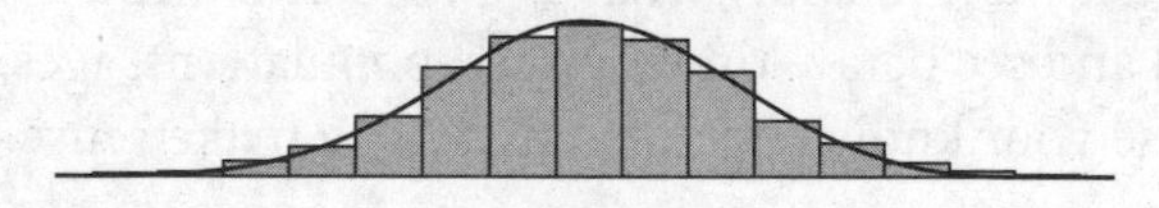

Figure 5–3. Histogram with Approximating Normal Curve

Measures of Central Tendency and Standard Deviation for the Normal Distribution

The ability to approximate data using the normal distribution makes it a very powerful and useful statistical tool. When you encounter real-world data that is approximately normal, you can model it with a continuous normal distribution and then apply everything you know about the normal distribution to your data. Since the normal distribution is symmetric and unimodal, it can be understood with just a few numbers. The mean, median, and mode are all the same; that is, a single center of the data splits the data into two parts that are mirror reflections of each other. This is one reason why normal distributions are easy to work with.

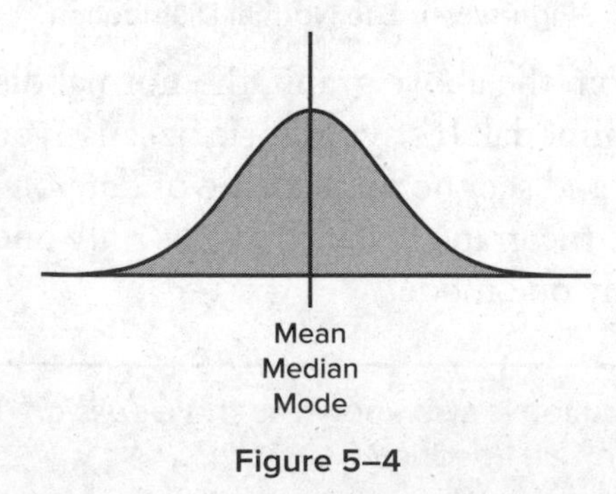

Figure 5–4

Here's an example that shows just how painless working with the normal distribution can be.

Example 1:

The SAT is a test used by many U.S. colleges for admissions purposes. Scores on the SAT range from 400 to 1600 and are approximately normally distributed with a population mean of $\mu = 1000$.

1+2=3 MATH TALK!

The symbol μ is the Greek letter mu (pronounced "mew") and is often used as the symbol for the mean of a data set, especially a population.

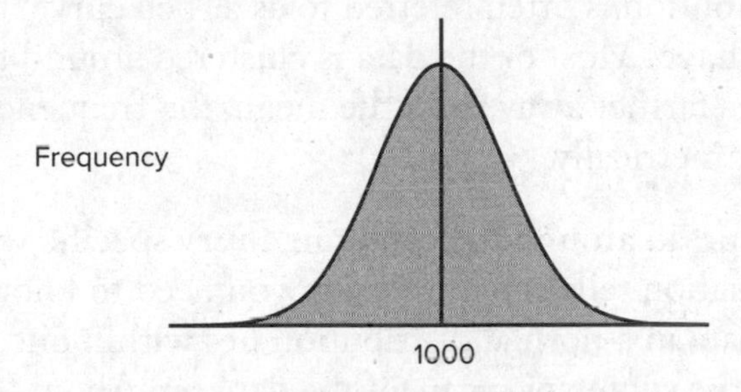

Figure 5–5. All SAT Scores

Since the mean SAT score is 1000, one way to think of an "average" test taker is as a student who scored a 1000 on the SAT. The mean, or average, is what you'd get if you added up all the test scores of all the test takers and then divided that sum by the total number of test takers.

Since the normal distribution is symmetric and unimodal, the median is the same as the mean, so the median of this data set is also 1000. This tells you that half of the test takers scored above 1000 and half scored below 1000.

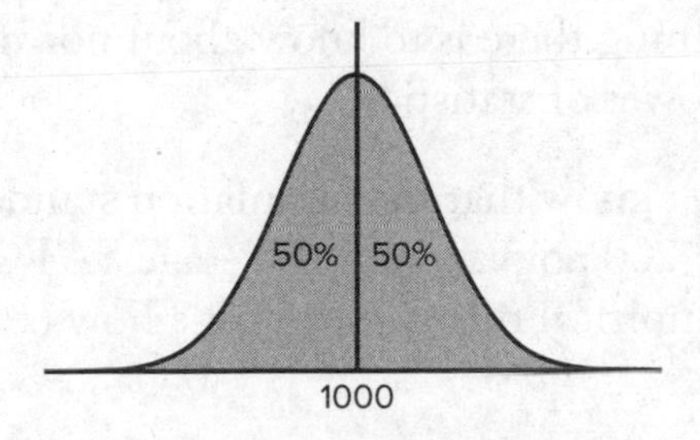

Figure 5–6. All SAT Scores

The mode of this data set is also 1000, meaning that the overall most frequent score on the SAT is around 1000.

REMINDER

Even when a known distribution like the normal distribution doesn't match real-world data exactly, if it's a good approximation, the known distribution can still be used to model and analyze the data.

The way that data is dispersed, or spread, about the center is a very important characteristic of the normal distribution. The graph of the normal distribution is often referred to as a "bell curve" because of its bell-like shape. Most of the data is clustered around the center, and for values farther away from the mean, the frequency of the data drops off symmetrically.

The data is spread around the center in a very specific way, and the standard deviation tells you everything you need to know. About 68% of the data in a normal distribution lies within one standard deviation of the center, or mean, of the data set. About 95% of the data lies within two standard deviations of the mean, and around 99.7% of the data lies within three standard deviations of the mean. This is often referred to as the *empirical rule*, or the 68-95-99.7 rule, for normal distributions.

The empirical rule tells you that most normally distributed data is close to the average. Since 95% of the data is within two standard deviations of the mean, only 5% of the data is farther than two standard deviations from the mean. Thanks to the empirical rule, with only two numbers (the mean and the standard deviation), you can know most everything there is to know about normally distributed data. That's the power of statistics!

For example, if you know that the population standard deviation of SAT scores is $\sigma = 200$ points and the average SAT score is $\mu = 1000$, you can use the empirical rule to determine how many scores occur in certain ranges.

1+2=3 MATH TALK!

The symbol σ is the Greek lowercase letter sigma and is often used as the symbol for the standard deviation of a data set, especially a population.

In the SAT data, one standard deviation above the mean would be 1000 points + 200 points = 1200 points, and one standard deviation below the mean would be 1000 points − 200 points = 800 points. By the empirical rule, around 68% of test takers scored between 800 and 1200 points.

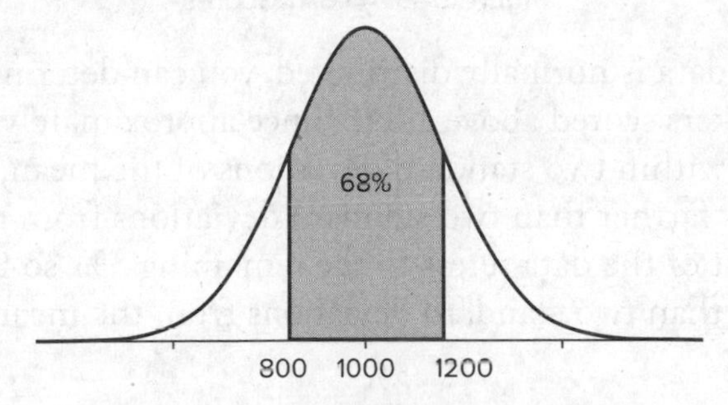

Figure 5–7. All SAT Scores

Also, by the empirical rule, around 95% of the data lies within two standard deviations of the mean. Since one standard deviation is 200 points, two standard deviations are 400 points. Thus, two standard deviations above the mean is 1000 points + 400 points = 1400 points, and two standard deviations below the mean is 1000 points − 400 points = 600 points. So, around 95% of test takers scored between 600 and 1400 points on the SAT.

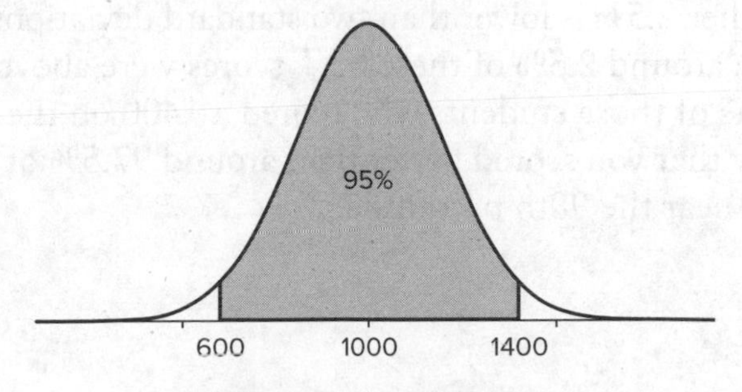

Figure 5–8. All SAT Scores

Notice the unshaded part of the graph at that right end. Those are the SAT scores that are above 1400, or more than two standard deviations above the mean.

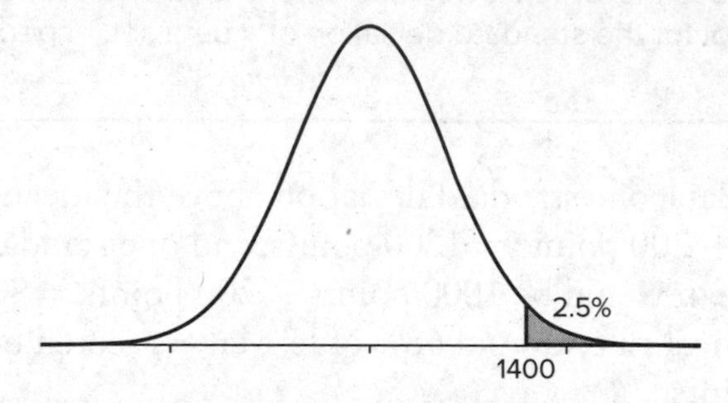

Figure 5–9. All SAT Scores

Because the data is normally distributed, you can determine how many test takers scored above 1400. Since approximately 95% of the data lies within two standard deviations of the mean, the rest of the data is farther than two standard deviations from the mean. Here, the rest of the data refers to the remaining 5%, so 5% of the data is more than two standard deviations from the mean.

PAINLESS TIP

If you know that X% of the data lies in a certain region, then the remaining region(s) must contain the remaining 100% – X% of the data. This is sometimes known as the *complement rule*.

Since the normal distribution is symmetric, half of that 5%, or 2.5% of the data, is higher than two standard deviations above the mean, while the other 2.5% is lower than two standard deviations below the mean. Thus, around 2.5% of these SAT scores were above 1400. If you were one of those students who scored a 1400 on the SAT, you would know that you scored better than around 97.5% of test takers, putting you near the 98th percentile.

The population standard deviation, σ, is the standard deviation of the entire data set, whereas the sample standard deviation, S, is used when looking at samples of a population. Similarly, the population mean is denoted by the Greek letter μ, while the sample mean is usually denoted by $\bar{x}$.

Here's another example of how you can use only two numbers (the mean and the standard deviation) to analyze normally distributed data.

Example 2:

Human height is another kind of data that can be approximated with the normal distribution. Assume that the heights of adult American men are normally distributed with a mean of 70 inches and a standard deviation of 3 inches. How does an American man who is 6 feet, 1 inch tall compare to the rest of the men in America?

Well, 6 feet, 1 inch is 73 inches, which is above the average height of 70 inches. So a man who is 6 feet, 1 inch tall is taller than average adult American man. Is he unusually tall? The properties of the normal distribution allow you to answer this question. Since the standard deviation of the data is 3 inches, his height is one standard deviation above the mean, which puts him roughly here in the data set:

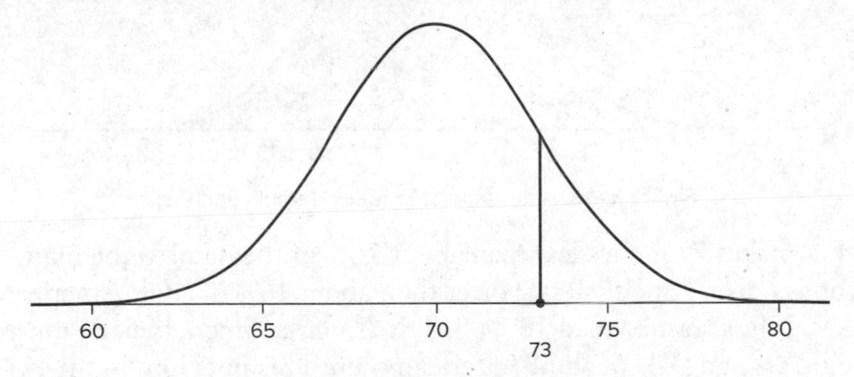

Figure 5–10. Adult American Male Heights (inches)

You can see that he's above (to the right of) the mean, and since the median is equal to the mean, at least 50% of the graph is below (to the left of) him. The properties of the normal distribution, however, allow you to say much more. By the empirical rule, 68% of the data in a normal distribution lies within one standard deviation of the mean. Since the data is symmetric, that's 34% on each side of the median.

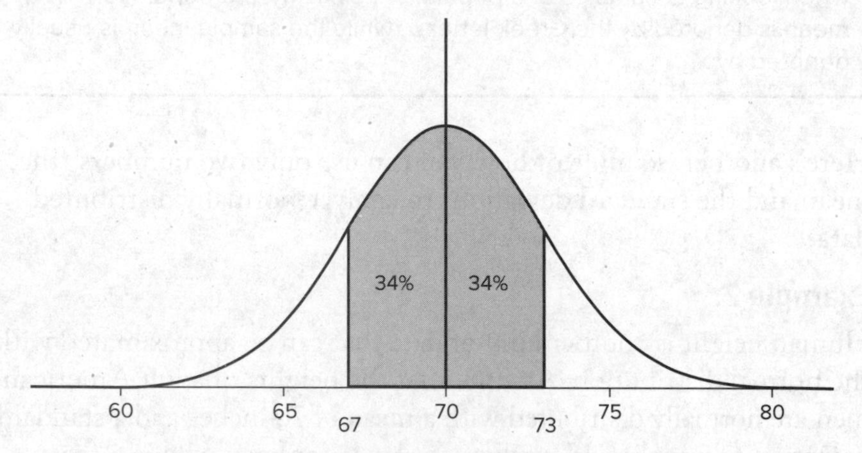

Figure 5–11. Adult American Male Heights (inches)

Of the 50% of the data that lies above the median, 34% is less than one standard deviation above it, which means the other 16% of the data is more than one standard deviation above the center line.

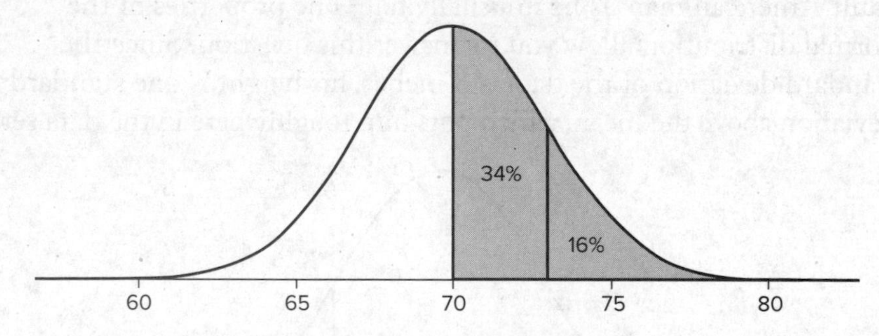

Figure 5–12. Adult American Male Heights (inches)

That makes 73 inches less than about 16% of the data, so the man who is 6 feet, 1 inch tall is shorter than about 16% of adult American men. If he's shorter than 16% of adult American men, then he must be taller than 84% of adult American men. This puts him in the 84th percentile for height: above average but not especially unusual.

Now, what about a man who is 6 feet, 4 inches tall? How does he compare to the rest of the men in America? That's 76 inches, which is two standard deviations above the mean. You can use the empirical rule again to compare this height to the rest of the population. If 95% of the data lies within two standard deviations of the mean, then by symmetry, the remaining 5% is split evenly between the region that is higher than two standard deviations above the mean and the region that is lower than two standard deviations below the mean.

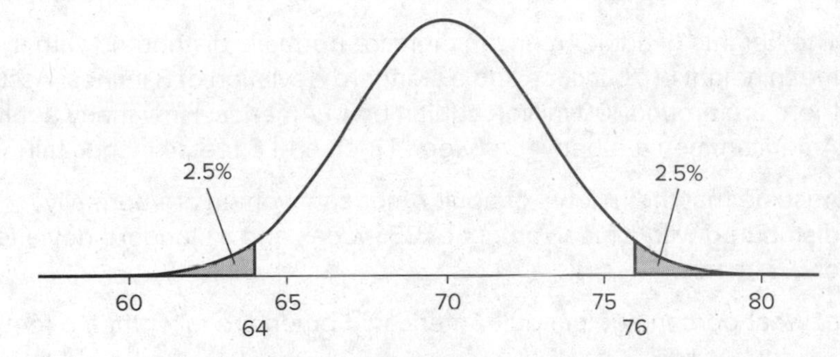

Figure 5–13. Adult American Male Heights (inches)

That makes someone who is 6 feet, 4 inches taller than 97.5% of adult American men. Notice the dramatic difference: one standard deviation above the mean is higher than 84% of the data, but two standard deviations above the mean are higher than 97.5% of the data. This is because in a normal distribution, most of the data is clustered around the mean.

In addition to analyzing individual data points, you can use knowledge of the normal distribution to better understand the data set as a whole. For example, there are around 100 million adult American men. Assuming the population mean height is $\mu = 70$ inches and the population standard deviation is $\sigma = 3$ inches, you can say that around 68 million adult American men are between 5 feet, 7 inches and 6 feet, 1 inch tall (because 68% of the data in a normal distribution should fall within one standard deviation of the mean). A similar analysis shows that there are only around 150,000 adult American men who are 6 feet, 7 inches or taller (since in a normal distribution, less than 0.15% of the data should be three standard deviations above the mean or higher). Knowing about height distributions in the population would be helpful if you ran a clothing business or designed homes or airplane seats, and there are many more real-world applications of the normal distribution to data.

BRAIN TICKLERS Set #14

1. The heights of adult American men are normally distributed with a mean height of 70 inches and a standard deviation of 3 inches. What percentage of adult American men are less than 5 feet, 7 inches tall?
2. The heights of adult American men are normally distributed with a mean height of 70 inches and a standard deviation of 3 inches. Assume there are around 100 million adult men in America. How many adult American men are between 6 feet, 1 inch and 6 feet, 4 inches tall?
3. Assume that the heights of adult American women are normally distributed with a mean height of 65 inches and a standard deviation of 2.5 inches.
   a. What percentage of adult American women are taller than 5 feet, 5 inches tall?
   b. If there are roughly 110 million adult women in the U.S., how many women are less than 5 feet tall?
   c. What percentage of adult American women are taller than the average height for an adult American man (which is 70 inches)?

(Answers are on page 127.)

Comparing Data Using z-Scores

Comparing Different Normal Distributions

Normally distributed data is easy to work with and understand as a whole. In a normal distribution, you can compare individual data points to the rest of the data set, but it's also possible to compare data from two entirely different normal distributions, as seen in the next example.

Example 3:

Many kinds of test scores are distributed in a way that can be approximated with the normal distribution. For example, suppose a researcher gave 1,000 people a memory test, and the results were approximately normal, with an average score of 25 and a standard deviation of 5.

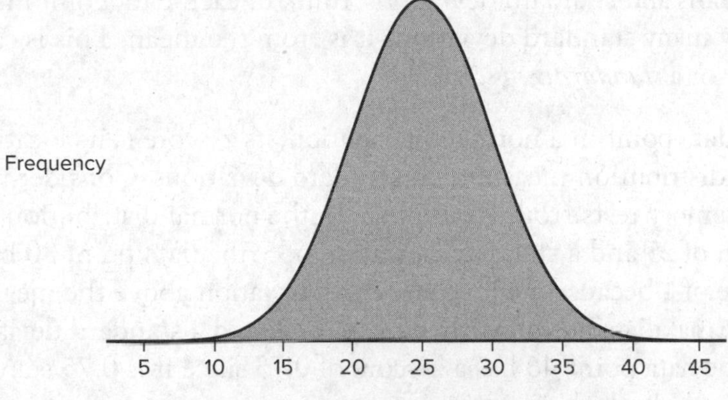

Figure 5–14. Scores on a Memory Test

If you took this test and scored a 30, that would mean you have a pretty good memory. That score would put you one standard deviation above the mean, and as a consequence of the empirical rule, your score would be higher than around 84% of the people who took the test.

Now, suppose your friend took a different memory test from a different researcher and scored a 46. Your friend might brag that she has a better memory than you because she got a higher score, but in general, it would be difficult to compare your results. After all, you and your friend took different tests. However, if the scores of both tests are normally distributed, comparing scores is easy.

Suppose the test your friend took had an average score of 40 and a standard deviation of 8. Your friend's score of 46 may be higher than your score of 30, but instead of comparing raw scores, you should compare where your scores lie in terms of standard deviations. Your score of 30 is a full standard deviation above the average on your test, while your friend's score of 46 is less than one standard deviation higher than the average on her test. Comparing your results in this way, you did better on your test than your friend did on hers; thus, this comparison is possible even though you took completely different tests.

z-Scores and Standard Normal Distributions

To compare two different, normally distributed data sets with different means and standard deviations, think of each data point in terms of how many standard deviations it is from the mean. This is called a *z-score* or a *standardized score.*

For a data point in a normal distribution, its z-score is its location in the distribution measured in standard deviations. Consider the two memory tests from Example 3. In the normal distribution with a mean of 25 and a standard deviation of 5, the data point 30 has a z-score of 1 because it is one standard deviation above the mean. In the normal distribution with a mean of 40 and a standard deviation of 8, the data point 46 has a z-score of 0.75 since it is 0.75 standard deviations above the mean.

The 30 and the 46 in this memory test example cannot be directly compared because they are data points from different distributions, but their z-scores can be directly compared. You can think of the z-scores as data points in the same distribution: the standard normal distribution. The *standard normal distribution* is a normal distribution with a mean of 0 and a standard deviation of 1.

You can turn any normal distribution with mean μ and standard deviation σ into the standard normal distribution using the following painless steps.

Step 1: Subtract the mean, μ, from every point in the data set.

Step 2: Divide every resulting number by the standard deviation, σ.

By subtracting the mean from every data point in the data set, you turn the original data set into a new data set with a mean of 0. (You can show this for any discrete data set using the formula for mean and some algebra.) Graphically, you can think of this as just sliding the graph over so that its center is now at 0 instead of at μ.

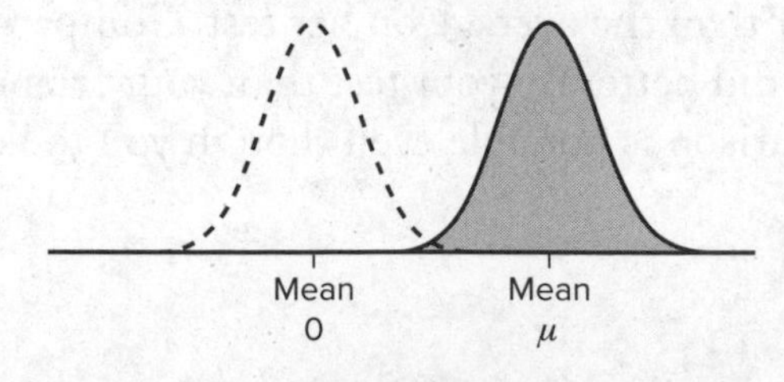

Figure 5–15

Since every data point is moved by the same amount, the center of the distribution is changed but everything else about the shape of the data remains the same.

The result of dividing the new data (with a mean of 0) by the standard deviation has the effect of making the standard deviation of the new data set exactly 1. (This can also be shown for any discrete data set using the formula for standard deviation and some algebra.) Graphically, this has the effect of compressing (or possibly stretching) the data set around its new center of 0.

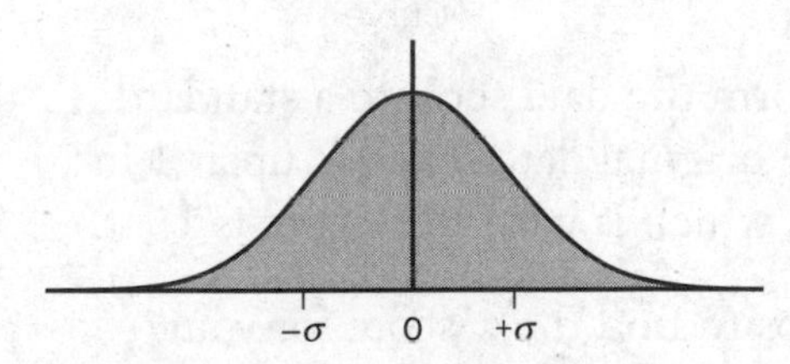

Figure 5–16. Normal Distribution with Mean 0 and Standard Deviation σ

Figure 5–17. Normal Distribution with Mean 0 and Standard Deviation 1

The standard normal distribution is a normal distribution with a mean of 0 and a standard deviation of 1. So the empirical rule says that 68% of the data in the standard normal distribution lies between −1 and 1; roughly 95% of the data lies between −2 and 2; and over 99% of the data lies between −3 and 3.

The standard normal distribution is where z-scores come from. Think of the data point 30 in the normal distribution with a mean of 25 and a standard deviation of 5. If you subtract the mean from the data point, you get $30 - 25 = 5$. Now, divide by the standard deviation: $\frac{5}{5} = 1$. Thus, the data point of 30 in a normal distribution with $\mu = 25$ and $\sigma = 5$ has a z-score of 1.

In summary, to find the z-score of an individual data point, x, follow these two painless steps.

Step 1: Subtract the mean, μ, from x: $x - \mu$.

Step 2: Divide that new number by the standard deviation, σ: $\frac{x - \mu}{\sigma}$. This is the z-score.

One way to visualize this process is to imagine transforming the data set into the standard normal distribution and following where the individual data point goes.

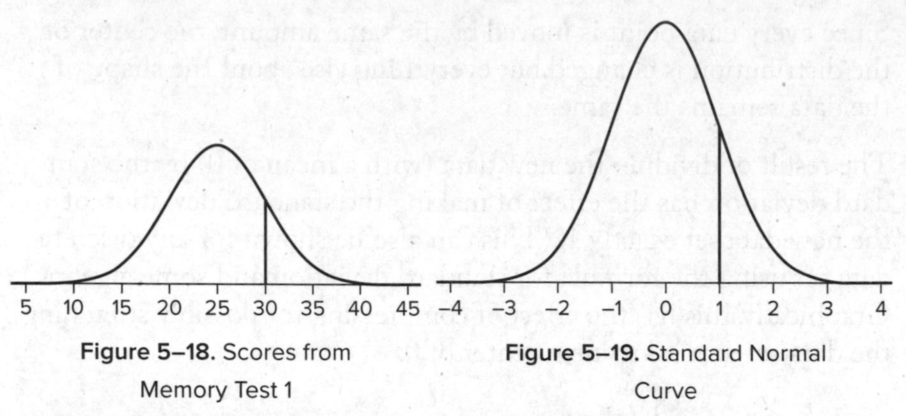

Figure 5–18. Scores from Memory Test 1

Figure 5–19. Standard Normal Curve

In this example, when you transform the data set into a standard normal distribution, the 30 in the original data set ends up at 1 in the standard normal distribution, which is why its z-score is 1.

Now, you can apply the same steps to find the z-score for your friend's test score of 46 from the second memory test, which has a mean of 40 and a standard deviation of 8:

$$z = \frac{x - \mu}{\sigma} = \frac{46 - 40}{8} = \frac{6}{8} = 0.75$$

So, the 46 in the original data set has a z-score of 0.75. Again, it may help to visualize this by following where the 46 goes once the original data set is transformed into the standard normal distribution.

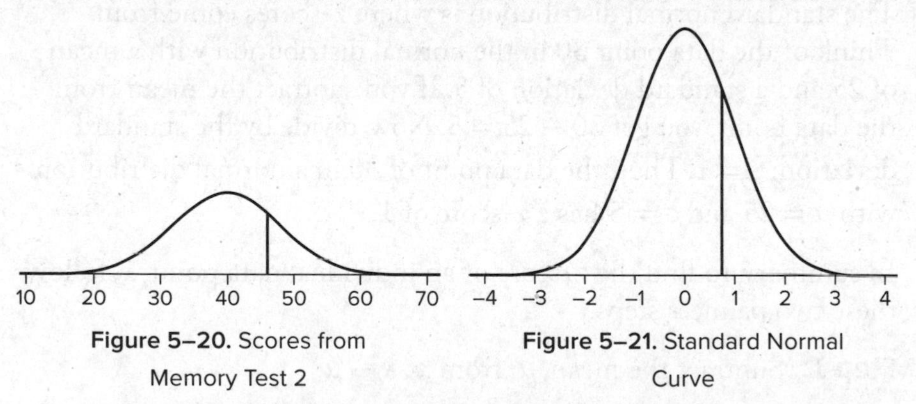

Figure 5–20. Scores from Memory Test 2

Figure 5–21. Standard Normal Curve

What's amazing about this process is that, although your two data points start off in two entirely different data sets with different means and different standard deviations, they end up in the same data set: the standard normal distribution. Furthermore, now that they're in the same data set, you can compare them directly.

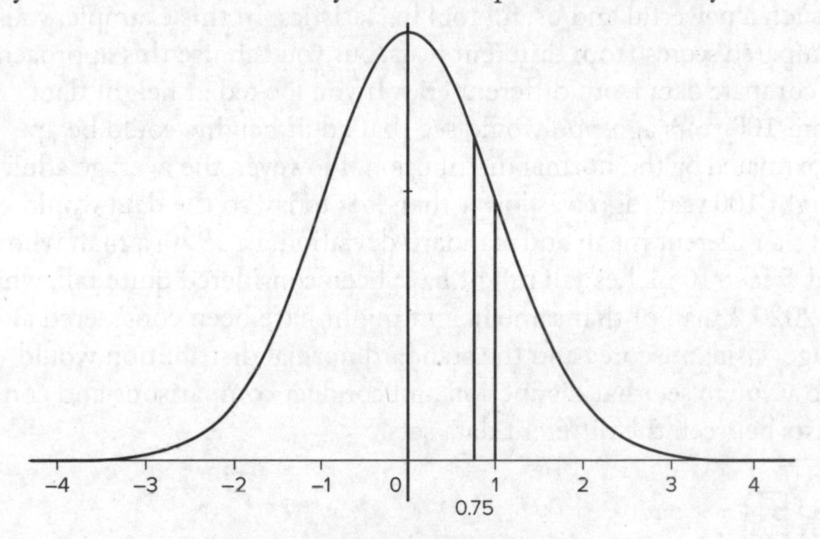

Figure 5–22. Standard Normal Curve

Instead of thinking about a data point of 30 in a data set with a mean of 25 and a standard deviation of 5, just think of the z-score of 1. Instead of a data point of 46 from a data set with a mean of 40 and a standard deviation of 8, just think of a z-score of 0.75. These z-scores are directly comparable because they both represent a location in the same data set: a standard normal distribution with a mean of 0 and a standard deviation of 1. Thus, even though your friend's score might initially look higher, you know you did better on your test than she did on hers.

So, whose memory is better? It's hard to say for sure. If the different tests really do measure the same kind of ability, just on different scales, then it seems like you've come out on top. However, it's hard to know exactly what these two tests really measure. Although z-scores are easy to compare, real-world questions are rarely easy to answer.

Statistics can provide useful information about data, but data alone never tells the whole story. It's important to understand the data in context to really understand a situation.

The ability to directly compare different data sets by transforming the data into z-scores is another reason why the normal distribution is such a powerful and useful tool in statistics. In this example, you compared scores from different tests, but you can use this approach to compare data from different eras. If you looked at height data from 100 years ago, you would see that adult heights could be approximated by the normal distribution. However, the average adult height 100 years ago was lower than it is today, so the data would have a different mean and standard deviation. In 1920, a man who was 5 feet, 10 inches tall might have been considered quite tall, while in 2020, a man of that same height might have been considered average. Using z-scores and the standard normal distribution would allow you to see that distinction and conduct comparisons and contrasts between the different data sets.

BRAIN TICKLERS Set #15

1. SAT scores are normally distributed with a mean of 1000 and a standard deviation of 200. Find the z-score associated with each of the following SAT scores.

   a. 1000

   b. 950

   c. 1550

   d. 850

2. The z-score associated with a man's height is –1. Assuming men's heights are normally distributed with a mean of 70 inches and a standard deviation of 3 inches, how tall is this man?

3. Another friend took a different memory test and scored a 120 on that test (whose scores are normally distributed with a mean of 100 and a standard deviation of 15). Is this better or worse than scoring a 30 on the first memory test (with a mean of 25 and a standard deviation of 5)?

(Answers are on page 127.)

Standard Normal Tables

How to Read the Standard Normal Table

Every normal distribution can be understood using z-scores. Because of this, statisticians created a *standard normal table.* This is a table that lets you determine exactly where every z-score lies in a standard normal distribution.

Here's what an excerpt of a standard normal table looks like:

z	0.00	0.05
−0.5	0.309	0.291
−0.4	0.345	0.326
−0.3	0.382	0.363
−0.2	0.421	0.401
−0.1	0.460	0.440
−0.0	0.500	0.480
0.0	0.500	0.520
0.1	0.540	0.560
0.2	0.579	0.599
0.3	0.618	0.637
0.4	0.655	0.674
0.5	0.692	0.709

The entries in this standard normal table represent the percentage of data that is less than the given z-score. In other words, the entry in the table gives you the area under the curve to the left of your z-score.

> In Chapter 2, you learned that the graph of a cumulative distribution function (CDF) shows you the percentage of data that is less than or equal to a specific value. That's exactly what entries in the standard normal table tell you because the standard normal table is the CDF of the standard normal distribution given in table form (instead of as a graph).

For example, find the row with a 0.3 in the first column. The number to the right of that in the second column, 0.618, means that, in a standard normal distribution, 61.8% of the data is below 0.3.

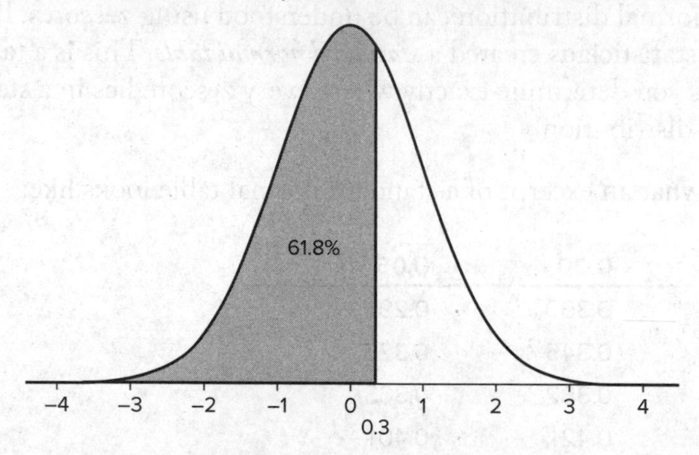

Figure 5–23. Standard Normal Curve

The additional columns in the standard normal table give you additional precision. The 0.637 (to the right of the 0.618 in the 0.3 row) means that 63.7% of the data lies below 0.35 on the standard normal distribution.

PAINLESS TIP

The leftmost column in the standard normal table gives you the initial digits of a z-score. As you move to the right in the table, you add the amount in the column header to those initial digits. For example, in the 0.3 row, the entry in the 0.00 column corresponds to the z-score of 0.3 + 0.00 = 0.30, whereas the entry in the 0.05 column corresponds to the z-score of 0.3 + 0.05 = 0.35.

Applying the Standard Normal Table

The standard normal table can be used to analyze and understand real-world data in different ways. Examples 4, 5, and 6 show you the most common applications of the standard normal table to normally distributed data.

Example 4:

Recall the SAT scores data from Example 1, earlier in this chapter. SAT scores are normally distributed with a mean of 1000 and a standard deviation of 200. Now, suppose someone scores a 1050 on the SAT. How does that score compare to the test takers in general?

To answer this question, first compute the z-score associated with 1050. To do this, follow the painless steps for finding a z-score. First, subtract the mean of 1000 from the SAT score of 1050, and then divide that number by the standard deviation of 200:

$$z = \frac{1050 - 1000}{200} = \frac{50}{200} = \frac{1}{4} = 0.25$$

Thus, a 1050 SAT score has a z-score of 0.25.

Now, look up 0.25 on the standard normal table. To do that, first find the 0.2 in the first column, and then move across to get more precision with more significant digits. After finding the 0.2 row in the first column, to locate the z-score of 0.25, move across to the 0.05 column.

z	0.00	0.05
−0.5	0.309	0.291
−0.4	0.345	0.326
−0.3	0.382	0.363
−0.2	0.421	0.401
−0.1	0.460	0.440
−0.0	0.500	0.480
0.0	0.500	0.520
0.1	0.540	0.560
0.2	0.579	0.599
0.3	0.618	0.637
0.4	0.655	0.674
0.5	0.692	0.709

Figure 5–24

The entry in the standard normal table for a z-score of 0.25 is 0.599, which means that 59.9% of the data is less than your data point. Thus, in this case, around 59.9% of SAT scores are less than 1050, which puts a score of 1050 on the SAT roughly in the 60th percentile.

Notice that, by the complement rule, you also know that around 100% – 59.9% = 40.1% of the data is above a z-score of 0.25, or an SAT score of 1050.

CAUTION—Major Mistake Territory!

The standard normal table only tells you how much data lies to the left of your data point. If you want to know how much data lies to the right, or between two values, the number from the table is *not* the answer. In those cases, you would have to use the complement rule, or perform some subtraction, to get the answer you want.

Example 5:

Now, what about an SAT score of 900? How does that score compare to test takers in general if the scores are normally distributed with a mean of 1000 and a standard deviation of 200?

To find the answer, first compute the z-score associated with 900:

$$z = \frac{900 - 1000}{200} = \frac{-100}{200} = -\frac{1}{2} = -0.5$$

Based on this calculation, an SAT score of 900 has a z-score of –0.5. On the standard normal table, the entry for a z-score of –0.5 is 0.309.

z	0.00	0.05
–0.5	0.309	0.291
–0.4	0.345	0.326
–0.3	0.382	0.363
–0.2	0.421	0.401
–0.1	0.460	0.440
–0.0	0.500	0.480
0.0	0.500	0.520
0.1	0.540	0.560
0.2	0.579	0.599
0.3	0.618	0.637
0.4	0.655	0.674
0.5	0.692	0.709

Figure 5–25

That means that roughly 30.9% of SAT scores are less than 900, and by the complement rule, about 69.1% of the data is above a score of 900.

Example 6:

Finally, suppose you have two SAT scores, 1050 and 900, and you want to determine how much data lies between those values (still assuming the data is normally distributed, the mean is 1000, and the standard deviation is 200). To do this, you can use the standard normal table to compute how much data lies in this range. For example, you just learned that for a z-score of 0.25, 59.9% of the data lies to the left, and for a z-score of −0.5, 30.9% of the data lies to the left of that.

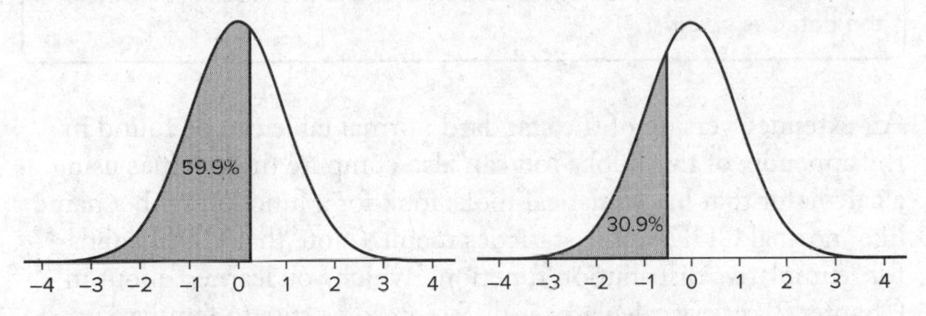

Figure 5–26. Standard Normal Curve **Figure 5–27.** Standard Normal Curve

You can compute how much of the data lies between z-scores of −0.5 and 0.25 by subtracting the smaller amount from the larger amount. Here you get 0.599 − 0.309 = 0.290; thus, 29% of the data lies between z-scores of −0.5 and 0.25.

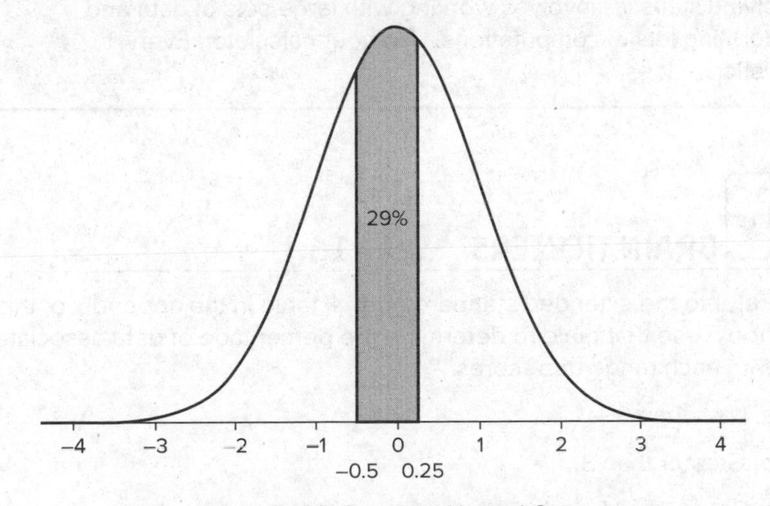

Figure 5–28. Standard Normal Curve

So, in a normal distribution, 29% of the data is higher than 0.5 standard deviations below the mean but less than 0.25 standard deviations above the mean. In the context of SAT scores, this means that 29% of SAT scores lie between 900 and 1050.

PAINLESS TIP

The standard normal table contains a row for both 0.0 and −0.0. This allows you to find z-scores that are both slightly above and slightly below 0. Also, notice that the entry in the table for 0.0 is 0.5. That's because in the standard normal distribution, 0 is the median, so 50% of the data lies below it.

An extended version of the standard normal table can be found in the appendix of this book. You can also compute these values using a calculator that has statistical tools: look for a function with a name like "normal CDF" in the statistics menu. (Note that CDF stands for "cumulative distribution function," which you learned about in Chapter 2). Every calculator is different, so be sure to familiarize yourself with your calculator's functions.

REMINDER

Applying statistics involves working with large sets of data and performing lots of computations. Use your calculator! Every statistician does.

BRAIN TICKLERS Set #16

1. Refer to the extended standard normal table in the appendix of this book. Use that table to determine the percentage of data associated with each range of z-scores.

   a. Less than 0

   b. Greater than 3

   c. Between −1.5 and 1.5

2. Assuming that the heights of adult American men are normally distributed with a mean of 70 inches and a standard deviation of 3 inches, approximately how many of the 100 million adult men in the U.S. are taller than 6 feet?
3. In healthy adults, systolic blood pressure (the pressure caused by the heart contracting and pushing out blood) is normally distributed with a mean of $\mu = 112$ mmHg and a standard deviation of $\sigma = 10$ mmHg. (mmHg stands for millimeters of mercury, a measure of pressure.) What percentage of blood pressure readings are between 107 and 117 mmHg?

(Answers are on page 127.)

When to Use the Normal Distribution

The normal distribution is very useful because so much data in the world is normally distributed and the properties of this distribution make it easy to work with and understand. However, it's important to remember that not all data is normally distributed. When you encounter new data, you can't just assume the data is normally distributed. You have to think about whether such a justification is warranted. If you use the normal distribution when it shouldn't be used, you will end up drawing bad conclusions and making bad decisions.

For example, scores on a memory test may be normally distributed if the test is given to a wide variety of different people. However, what if you gave the same memory test to a group of chess grand masters? To be successful at chess, a person needs to be able to remember lots of different positions, combinations, and strategies. To be a chess grand master, a person probably needs to have a good memory, so a chess grand master would likely score high on a memory test. If you looked at the scores of chess grand masters compared with those of the entire population, they would probably be one or more standard deviations above the mean. So, if you gave a group of chess grand masters a memory test, the data set of their scores probably wouldn't be normally distributed; their scores would look more like the tail of the normal distribution.

The same is true of SAT scores. Over the entire testing population, scores are normally distributed. On the other hand, if you looked only at the scores of students at an elite college like Stanford

University, those scores wouldn't be normally distributed. To get into a school like Stanford, a student probably needs a high SAT score, so most of that data would come from the tail of the distribution. If you looked at that data by itself, the normal distribution would not approximate it accurately.

Think carefully about your data when deciding how to approximate it. For example, it is true that the heights of adult American men can be approximated with the normal distribution. However, the heights of all Americans is a very different data set. The age of Americans is not uniformly distributed. Over 10% of Americans are under 10 years old. Since children are often shorter than adults, if you added them to the data set of adult heights, they would skew the data and pull the mean and median down.

You also have to be careful about comparisons across subgroups. For example, the heights of adult women can be approximated with the normal distribution, just like the heights of adult men. However, the mean will be different since women tend to be, on average, shorter than men.

The more you learn about statistics, the more you'll see the normal distribution. Remember, though, that not all data is normally distributed. When deciding whether or not to use the normal distribution, think about the key features of the distribution. The normal distribution is symmetric, and most of the data is near the mean. If there's a good chance that something very unusual could happen—like a flood or an outbreak or a bankruptcy—then the normal distribution might not be the best choice to model your data.

Brain Ticklers—The Answers

Set #14, page 112

1. 16%
2. Around 13.5 million men
3. a. 50%

   b. Around 2.75 million

   c. Around 2.5%

Set #15, page 118

1. a. 0

   b. –0.25

   c. 2.75

   d. –0.75
2. 67 inches, or 5 feet, 7 inches tall
3. Better, since the 120 has a z-score of 1.33 and the 30 has a z-score of 1

Set #16, pages 124–125

1. a. 50%

   b. 0.13%

   c. 86.64%
2. Around 25.4 million
3. Around 38.3%

Chapter 6

The Fundamentals of Probability

Probability is the mathematics of likelihood. How likely are you to flip a coin three times in a row and see tails each time? How likely are you to roll two dice and get a sum greater than 10? How likely is it that it will rain tomorrow? Probability questions can range from very simple to very complex, but mathematical techniques can help you understand and answer them.

Probability is essential to statistics because many interesting statistical questions are inherently probability questions. How unusual is it for a man to be 6 feet, 5 inches tall? How unlikely would it be for a company's stock price to drop 50%? How likely is it that Candidate A will win the election? To understand these questions with statistics, you must first understand basic probability.

Basic Probability

Probability is a measure of how likely an event is to occur. An *event* could be anything from flipping tails with a fair coin to a stock price rising to a successful medical procedure. Events can be represented with variables: if T represents the event "seeing tails on a coin flip," you could write $P(T)$, or $Pr(T)$, to denote the probability of a coin flip coming up tails.

The probability of an event is always a number between 0 and 1. When flipping a fair coin, $P(T) = \frac{1}{2}$, which means that there is a 1 in 2 chance of flipping tails. You can also interpret probability as a percentage: since $P(T) = \frac{1}{2} = 0.5$, the probability of flipping tails is 50%.

1+2=3 MATH TALK!

The expression $P(E)$ is read "the probability of E," so an expression like $P(T) = \frac{1}{2}$ is read "the probability of T is one-half." Since T is the event that tails is flipped, you can also read this as "the probability of flipping tails is one-half."

If an event E has probability 1, that is, if $P(E) = 1$, then the event is guaranteed to happen: it is 100% likely. On the other hand, an event with probability 0, that is, $P(E) = 0$, is guaranteed not to happen.

PAINLESS TIP

The closer the probability of an event is to 1, the more likely it is to occur.

The field of probability is wide and complex, but the study of probability begins with a single painless formula.

The Fundamental Probability Formula

Here is the fundamental probability formula:

$$P(\text{Event}) = \frac{\text{number of favorable outcomes}}{\text{number of possible outcomes}}$$

This formula applies to *discrete* events with *equally likely outcomes*. A discrete event has a finite number of outcomes, and when each outcome has the same probability of occurring, the outcomes are called "equally likely." For example, a fair coin flip has two equally likely outcomes: heads or tails. Rolling a fair six-sided die has six equally likely outcomes: the top face of the die could show 1, 2, 3, 4, 5, or 6. Randomly selecting a card from a standard deck has 52 equally like outcomes, as each of the 52 cards is equally likely to be drawn.

A favorable outcome is an outcome you are looking for in an event. If you want to compute the probability of flipping tails on a fair coin flip, then tails is the favorable outcome. Since there is one favorable outcome and two possible outcomes, by the fundamental probability formula, the probability of flipping tails is $\frac{1}{2}$. If you want to compute the probability of rolling a number greater than 4 on a fair six-sided die, then the favorable outcomes are 5 and 6. Since there are two favorable outcomes and six total outcomes, the probability is $\frac{2}{6}$, which can be simplified to $\frac{1}{3}$.

The set of all possible outcomes of an experiment is called the *sample space*. For a coin flip, the sample space is the set {Heads, Tails}. For a fair, six-sided die roll, the sample space is $\{1, 2, 3, 4, 5, 6\}$. An event can be thought of as a subset of the sample space. Flipping tails corresponds to the subset {Tails}; rolling a number greater than 4 corresponds to the subset $\{5, 6\}$. With this knowledge, you can use the fundamental probability formula to compute the probability of an event in a discrete sample space with equally likely outcomes. To do so, just follow these painless steps.

Step 1: Determine the size of the sample space that corresponds to your experiment.

Step 2: Determine the size of the subset that corresponds to your event.

Step 3: Divide the size of the subset (Step 2) by the size of the sample space (Step 1).

Here's a quick example to show you how to apply these steps for computing probability.

Example 1:

If you draw one card from a standard deck of playing cards, what is the probability that the card drawn is a king?

The sample space of this event is a set that contains all 52 cards in the deck: {2 of clubs, 3 of clubs, . . .}. For the event "the card drawn is a king," the corresponding subset is {K of clubs, K of diamonds, K of hearts, K of spades}. Thus, the set of favorable outcomes has size 4, so the probability of drawing a king is $\frac{4}{52} = \frac{1}{13}$.

Sample Spaces and the Fundamental Counting Principle

Some experiments involve more complicated sample spaces than a single coin flip or die roll. Here's an example.

Example 2:

Imagine flipping a fair coin three times in a row and recording the sequence of flips. What are the outcomes of this experiment?

One way to visualize the outcomes, and thus the sample space, is with a tree diagram.

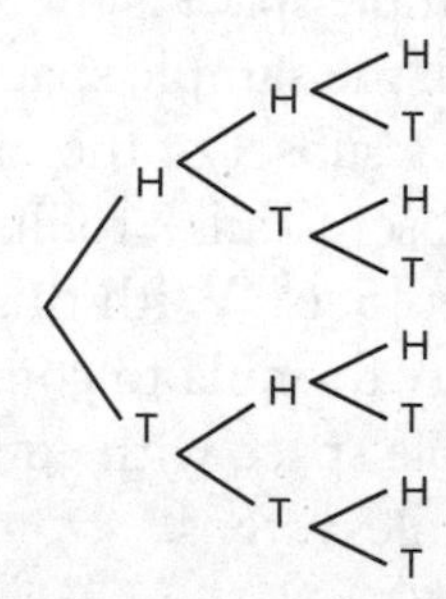

Figure 6–1. Tree Diagram for Three Coin Flips

Starting at the left, each point in the tree diagram where a branch splits into two indicates a coin flip. Branching up represents flipping heads, and branching down represents flipping tails. Notice how the three vertical levels, or columns, of the tree diagram correspond to the three coin flips.

Starting from the first coin flip on the left, you can follow a path through the tree diagram to end up at a sequence of heads and tails at the far right, as seen in the following figure.

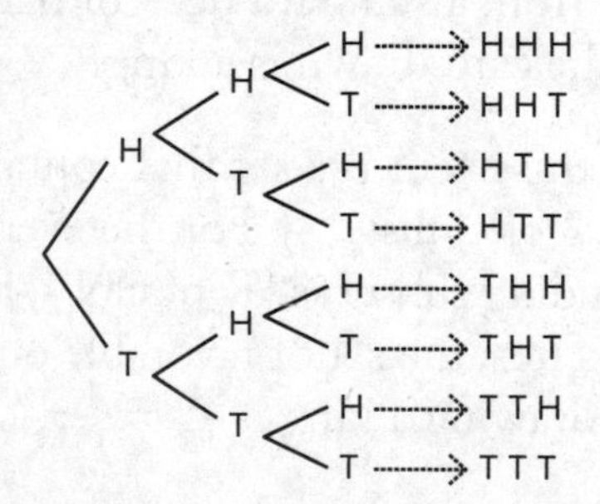

Figure 6–2. Tree Diagram for Three Coin Flips and Sample Space

At the top of the list is HHH, which corresponds to flipping heads three times in a row. The fourth entry on the list, HTT, corresponds to flipping heads, then tails, and finally tails again. The list of sequences at the far right represents all the possible outcomes, so that is the sample space for this experiment:

{HHH, HHT, HTH, HTT, THH, THT, TTH, TTT}

Sample Space of Three Consecutive Coin Flips

Now that you know the size of the sample space, you can compute probabilities of various outcomes. The probability of flipping three heads in a row is $\frac{1}{8}$ because there is only a single favorable outcome: HHH. The probability that the third flip is tails is $\frac{4}{8}$, or $\frac{1}{2}$, because there are four favorable outcomes: HHT, HTT, THT, and TTT. The probability that two of the three flips come up tails is $\frac{3}{8}$, which corresponds to three favorable outcomes: HTT, THT, and TTH.

CAUTION—Major Mistake Territory!

A common error in this situation is to think that the probability of getting two tails among the three flips is $\frac{1}{4}$ because there is one favorable outcome among the four possible outcomes: 0 tails, 1 tail, 2 tails, or 3 tails. However, this probability is not $\frac{1}{4}$ because these four outcomes are not equally likely (for example, there are multiple ways to get 2 tails but only one way to get exactly 0 tails). Therefore, the fundamental probability formula doesn't apply.

Tree diagrams can be helpful in understanding the structure of experiments and their sample spaces, but it isn't always feasible to construct a tree diagram. Imagine trying to make a tree diagram of an experiment that consists of 10 consecutive coin flips. It would be very large; thus, it is useful to have other methods for determining the size of a sample space.

An alternate method relies on an important mathematical fact called the fundamental counting principle. The *fundamental counting principle* applies when you have multiple choices to make. It says that if there are m options for your first choice and n options for your second choice, then there are a total of $m \times n$ options for the pair of choices. A classic example of the fundamental counting principle involves counting the number of outfits you can make from your wardrobe: if you have three different pairs of pants to choose from and four different shirts, you can make $3 \times 4 = 12$ different outfits.

Here's how you could use the fundamental counting principle to compute the size of the sample space in Example 2. When thinking about flipping a coin three times in a row, imagine each outcome as a "word" made up only of Hs and Ts, and each word is three letters long (like HHH, THT, or TTT). You can imagine building each of these words by filling in each of three blanks with either an H or a T.

____ ____ ____

Since you have two choices for each blank—either H or T—the fundamental counting principle tells you that there are $2 \times 2 \times 2 = 8$ total words you can make. Of course, this agrees with the results of the tree diagram. However, the real virtue of using the fundamental counting principle is that you can use it to easily compute the size of the sample space of an experiment involving any number of coin flips, as seen in the following example.

Example 3:

In a new experiment, you now flip a fair coin ten times in a row. How many possible outcomes are there in this experiment?

Each outcome would be a word of length 10, consisting of Hs and Ts. To build an outcome in the sample space, you would fill in each of 10 blanks with either an H or a T.

____ ____ ____ ____ ____ ____ ____ ____ ____ ____

Since you have 2 choices for each blank, the fundamental counting principle tells you that the number of possible outcomes is:

$$2 \times 2 \times 2 \times 2 \times 2 \times 2 \times 2 \times 2 \times 2 \times 2 = 2^{10} = 1{,}024$$

There are 1,024 possible outcomes in the sample space, each of which is a ten-letter word of Hs and Ts (such as HHHTHHTHHT or TTHHTTHHTT). Trying to make a tree diagram for this experiment would require a lot of time and paper, but the fundamental counting principle makes it painless. There are also other ways to visualize sample spaces, as seen in the next example.

Example 4:

Consider the experiment of rolling a pair of six-sided dice. How would you visualize this sample space?

A useful approach is to imagine each outcome as an ordered pair: so $(2, 5)$ would represent the outcome of rolling a 2 on the first die and a 5 on the second die. There are six possible outcomes for each die. So in building the sample space, you have 6 choices for the first coordinate of the ordered pair and 6 choices for the second coordinate. By the fundamental counting principle, there are $6 \times 6 = 36$ outcomes in the sample space. These outcomes can be visualized in a table.

(1,6)	(2,6)	(3,6)	(4,6)	(5,6)	(6,6)
(1,5)	(2,5)	(3,5)	(4,5)	(5,5)	(6,5)
(1,4)	(2,4)	(3,4)	(4,4)	(5,4)	(6,4)
(1,3)	(2,3)	(3,3)	(4,3)	(5,3)	(6,3)
(1,2)	(2,2)	(3,2)	(4,2)	(5,2)	(6,2)
(1,1)	(2,1)	(3,1)	(4,1)	(5,1)	(6,1)

Figure 6–3. Sample Space of a Pair of Six-Sided Dice

If the dice are fair, the 36 outcomes are all equally likely. You can use this table to help compute various probabilities about the rolls of two fair dice. For example, to compute the probability that the sum of the two dice is 5, you can just look through the table for all favorable outcomes. There are four outcomes where the dice sum to 5, namely $(1, 4), (2, 3), (3, 2)$, and $(4, 1)$, so the probability that the sum of the two dice is 5 is $\frac{4}{36} = \frac{1}{9}$. Similarly, the probability that the sum of the two dice is greater than 9 is $\frac{6}{36} = \frac{1}{6}$ because there are six favorable outcomes: $(4, 6), (5, 5), (5, 6)$ $(6, 4), (6, 5)$, and $(6, 6)$.

Sample Spaces and Data Sets

In many ways, a sample space is like a data set. Here's an example that shows you the connection between data and probability.

Example 5:

Suppose 50 people were asked to identify their favorite color, and the results were recorded in the following table.

Favorite Color	Count
Red	11
Blue	16
Green	7
Yellow	3
Purple	13

Now imagine picking a person at random from the group and asking them to name their favorite color. What is the probability that the randomly selected person's favorite color is green?

In this scenario, a favorable outcome is a person whose favorite color is green. There are seven such people, so there are seven favorable outcomes. There are 50 total people to choose from, so there are 50 possible outcomes. Thus, the probability of choosing a person whose favorite color is green is:

$$P(\text{Green}) = \frac{7}{50}$$

Likewise, the probability that a randomly selected person's favorite color is blue is $P(\text{Blue}) = \frac{16}{50}$, and the probability the favorite color is yellow is $P(\text{Yellow}) = \frac{3}{50}$.

PAINLESS TIP

The word *percent* means "out of 100," so if you can easily express a fraction with a denominator of 100, you can easily convert it to a percentage. For example, since $\frac{7}{50}$ is equivalent to $\frac{14}{100}$, $\frac{7}{50}$ is equivalent to 14%. Similarly, since $\frac{2}{5}$ is equivalent to $\frac{40}{100}$, it is equivalent to 40%.

This is an example of an experiment where the outcomes are not equally likely. The sample space of outcomes is the set {Red, Blue,

Green, Yellow, Purple}, but the probabilities of each outcome are different. When working with probability, it's very important to identify whether the outcomes are equally likely or not. It makes a big difference in how you approach problems and how you compute probabilities.

CAUTION—Major Mistake Territory!

A very common error in probability is assuming outcomes are equally likely when they are not. This kind of erroneous thinking can be seen in the naive attitude of someone who thinks that the chance it will rain tomorrow is 50% because there are only two possible outcomes: either it rains or it doesn't. This certainly isn't true if you live in the desert or in the tropics!

BRAIN TICKLERS Set #17

1. What is the probability of flipping tails on two consecutive tosses of a fair coin?
2. Suppose a fair coin is flipped three times. What is the probability that you see tails at least once?
3. Suppose you roll two fair six-sided dice.

   a. What is the probability that the sum of the dice is 3?

   b. What is the probability that both rolls are greater than or equal to 5?

   c. What is the probability of the most likely sum?
4. Suppose you choose one of the fifty U.S. states at random. What is the probability that the state's name begins with an A?
5. A group of students were surveyed about their favorite class; the data is shown in the following table.

Favorite Class	Count
Math	35
English	55
History	35
Physics	30
Computer Science	45

If a student is selected at random, what is the probability the student's favorite class is physics?

(Answers are on page 156.)

Properties of Probability

Here again is the table from Example 5 that shows the data of favorite colors. What is the probability that a randomly selected person's favorite color is red?

Favorite Color	Count
Red	11
Blue	16
Green	7
Yellow	3
Purple	13

Since there are 11 favorable outcomes (a person whose favorite color is red) and 50 total outcomes (the total number of people being selected from), you can use the fundamental probability formula to answer this question. If R is the event that the selected person's favorite color is red:

$$P(R) = \frac{11}{50}$$

Here's a related question: What is the probability that a randomly selected person's favorite color is not red? There are several ways to approach this problem: one way is by considering the *complement*.

Complements of Events

The complement was introduced in Chapter 5 when computing percentages associated with the normal distribution. If E is some set in the sample space, the complement of E is everything else: it is everything in the sample space that isn't in E.

The event R (that a randomly selected person's favorite color is red) is associated with a subset of the sample space. The complement of that event, denoted R^c, is everything else in the sample space. There are 11 people in the sample space whose favorite color is red, so the size of the subset associated with R is 11. Since there are 50 total people in the sample space, everything else has a size of $50 - 11 = 39$; thus, R^c has size 39. That means there are 39 people whose favorite color is something other than red, so the probability that a randomly selected person's favorite color is not red is $\frac{39}{50}$. You can write this as:

$$P(R^c) = \frac{39}{50}$$

1+2=3 MATH TALK!

In probability, for an event E, the related event E^c is the *complement* of E and is read "E complement." The event E^c consists of all the outcomes in the sample space that do not belong to E. In other words, for a set of outcomes E, the complement E^c is everything else. Sometimes E^c is read "not E" because it corresponds to the negation of the event E.

Notice that there are 11 people whose favorite color is red, 39 people whose favorite color is not red, and $11 + 39 = 50$ (the total number of people in the group). Since everyone in the group either has red as their favorite color or doesn't have red as their favorite color, this makes sense. This is related to the following equation:

$$P(R) + P(R^c) = 1$$

This equation works out because $P(R) + P(R^c) = \frac{11}{50} + \frac{39}{50} = \frac{11 + 39}{50} = \frac{50}{50} = 1$, but you can also interpret this in terms of outcomes. If you select a person at random from this group, there is a 100% chance that either red is the person's favorite color (the event R) or red is not the person's favorite color (the event R^c). One of these events must occur, so the probability that either R or R^c occurs is 1.

This is true in general for an event and its complement. For any event E, the following equation is always true:

$$P(E) + P(E^c) = 1$$

Since $P(E)$ is the probability that the event E occurs, then $P(E^c)$ is the probability that the event E does not occur. One of these events must occur, so $P(E) + P(E^c) = 1$.

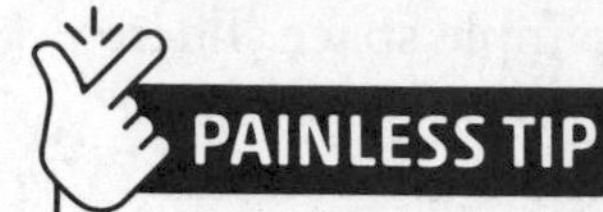

PAINLESS TIP

The equation $P(E) + P(E^c) = 1$ can be rearranged as $P(E^c) = 1 - P(E)$ to provide a formula for finding the probability of the complement of an event.

One of the reasons why this formula works is because events and their complements are disjoint. *Disjoint events* have no outcomes in common: there will never be an outcome that belongs to both an event and its complement. For example, a person's favorite color cannot be both red and not red. A coin flip cannot be both heads and not heads. A die roll cannot be both greater than 4 and less than or equal to 4. You'll revisit this concept and this important equation a little later in this chapter.

PAINLESS TIP

Disjoint events are also referred to as *mutually exclusive* events because if one happens, then the other cannot happen. For example, rolling an even number on a die and rolling an odd number on a die are mutually exclusive events because the result of a die roll must be either even or odd but cannot be both.

Events that are disjoint are easy to work with, but sometimes different events have outcomes in common. Working with the probabilities of these events is a littler trickier, as you'll see in the next section.

Compound Events

Sometimes in probability, an "event" is really multiple events happening simultaneously. To understand how to work with these *compound events*, consider the following example.

Example 6:

Imagine rolling a pair of fair six-sided dice (Die A and Die B). Consider the first coordinate in each ordered pair as the roll of Die A and the second coordinate as the roll of Die B. Here again is the sample space for this experiment.

(1,6) (2,6) (3,6) (4,6) (5,6) (6,6)
(1,5) (2,5) (3,5) (4,5) (5,5) (6,5)
(1,4) (2,4) (3,4) (4,4) (5,4) (6,4)
(1,3) (2,3) (3,3) (4,3) (5,3) (6,3)
(1,2) (2,2) (3,2) (4,2) (5,2) (6,2)
(1,1) (2,1) (3,1) (4,1) (5,1) (6,1)

Figure 6–4. Sample Space of a Pair of Six-Sided Dice

What is the probability that Die A shows a 1? This is easy enough to compute using the sample space and the fundamental probability formula. If E is the event that Die A shows a 1, then E corresponds to the subset $\{(1,1),(1,2),(1,3),(1,4),(1,5),(1,6)\}$ of the sample space. This subset has size 6, and since the sample space has size 36, you can use the fundamental probability formula: $P(E) = \frac{6}{36} = \frac{1}{6}$.

Similarly, if F is the event that Die B shows a 5, then F corresponds to the subset $\{(1,5),(2,5),(3,5),(4,5),(5,5),(6,5)\}$. In this case, $P(F) = \frac{6}{36} = \frac{1}{6}$.

Now, what is the probability that E and F both occur? This is an example of a *compound event*. The new event "E and F" is made up of the two individual events joined together by the word "and," so this new event requires both individual events to occur.

Since E is the event that Die A shows a 1 and F is the event that Die B shows a 5, there is only one outcome in the sample space that fit these criteria, $(1,5)$, which makes the probability of E and F equal to $\frac{1}{36}$. You can write this using probability notation as:

$$P(E \text{ and } F) = \frac{1}{36}$$

To compute the probability of a compound event, you reason logically about the sample space, or you can inspect the subsets associated with each individual event:

$$\text{Event } E: \quad \{(1,1),(1,2),(1,3),(1,4),(1,5),(1,6)\}$$

$$\text{Event } F: \quad \{(1,5),(2,5),(3,5),(4,5),(5,5),(6,5)\}$$

For event E to occur, one of the outcomes in the first subset must occur. For event F to occur, one of the outcomes in the second subset must occur. Notice that there is only one outcome that is in both subsets, namely $(1, 5)$. This outcome is the *intersection* of the two subsets, and in general, you can think of the compound event E and F as the *intersection* of the two events. The expression $E \cap F$ represents the intersection of the two events; that is, you can write $E \cap F$ instead of "E and F" in probability formulas:

$$P(E \cap F) = \frac{1}{36}$$

1+2=3 MATH TALK!

The upside-down U is the *intersection* symbol. The expression $A \cap B$ is read "A intersect B," and it refers to the set of all objects that are in both set A and set B.

Another important kind of compound event comes from joining events with the word "or," as seen in the next example.

Example 7:

Once again, imagine rolling two six-sided dice (Die A and Die B). Consider the first coordinate in each ordered pair as the roll of Die A and the second coordinate as the roll of Die B. E is the event that Die A shows a 1, so E corresponds to the subset $\{(1, 1), (1, 2), (1, 3), (1, 4), (1, 5), (1, 6)\}$ of the sample space. F is the event that Die B shows a 5, so F corresponds to the subset $\{(1, 5), (2, 5), (3, 5), (4, 5), (5, 5), (6, 5)\}$. What is the event "$E$ or F," and what is its probability?

This event occurs when Die A is a 1 *or* Die B is a 5. You could reason this out using the sample space, or you could again look at the subsets associated with those events.

Event E: $\{(1, 1), (1, 2), (1, 3), (1, 4), (1, 5), (1, 6)\}$

Event F: $\{(1, 5), (2, 5), (3, 5), (4, 5), (5, 5), (6, 5)\}$

The compound event E or F occurs when an outcome that is a part of E occurs or an outcome that is part of F occurs. This means the subset associated with the compound event E or F contains every outcome that is either in E or in F. This is the *union* of E and F and is written as $E \cup F$.

In this scenario, the event $E \cup F$ is associated with the subset $\{(1,1), (1,2), (1,3), (1,4), (1,5), (1,6), (2,5), (3,5), (4,5), (5,5), (6,5)\}$. These are all the individual outcomes that are either in E or in F. This subset has size 11, so the probability that Die A is a 1 or Die B is a 5 is $\frac{11}{36}$. You can write $P(E \cup F) = \frac{11}{36}$ or $P(E \text{ or } F) = \frac{11}{36}$.

1+2=3 MATH TALK!

You can read the compound event $E \cup F$ as "E union F" or "E or F."

A common error in probability is to expect that $P(E \text{ or } F) = P(E) + P(F)$. You can see that doesn't work out here. You already know that $P(E) = \frac{1}{6}$ and $P(F) = \frac{1}{6}$. Thus, $P(E) + P(F) = \frac{1}{6} + \frac{1}{6} = \frac{2}{6} = \frac{1}{3}$, which is not equal to $P(E \text{ or } F) = \frac{11}{36}$. Why doesn't this work? There are six outcomes where Die A is a 1 and six outcomes where Die B is a 5, so why aren't there 12 outcomes where Die A is a 1 or Die B is a 5?

The answer is that events E and F are not disjoint. The two events share an outcome in common, namely the outcome $(1,5)$. So if you try to count the outcomes of the compound event by adding the number of outcomes of each individual event, you end up counting $(1,5)$ twice. This explains the difference between $P(E) + P(F) = \frac{1}{3}$, which is equivalent to $\frac{12}{36}$, and $P(E \text{ or } F) = \frac{11}{36}$. In the former case, you've counted the outcome $(1,5)$ twice, giving you 12 favorable outcomes. In the latter case, you've counted $(1,5)$ only once, giving 11 favorable outcomes.

The Inclusion-Exclusion Principle

Although it is generally not true that $P(A \text{ or } B) = P(A) + P(B)$, this isn't far off. There's a special relationship called the *inclusion-exclusion principle* that relates these quantities, and it gives the following formula, which is true for any two sets A and B:

$$P(A \cup B) = P(A) + P(B) - P(A \cap B)$$

You can also phrase this relationship using the language of compound events:

$$P(A \text{ or } B) = P(A) + P(B) - P(A \text{ and } B)$$

You can verify the inclusion-exclusion formula for the events E and F from Example 7. Since $P(E) = \frac{1}{6}$, $P(F) = \frac{1}{6}$, $P(E \cup F) = \frac{11}{36}$, and $P(E \cap F) = \frac{1}{36}$, you can substitute as follows:

$$\begin{aligned} P(E \cup F) &= P(E) + P(F) - P(E \cap F) \\ \frac{11}{36} &= \frac{1}{6} + \frac{1}{6} - \frac{1}{36} \\ \frac{11}{36} &= \frac{6}{36} + \frac{6}{36} - \frac{1}{36} \\ \frac{11}{36} &= \frac{11}{36} \end{aligned}$$

You can see that the equation is indeed true.

The inclusion-exclusion formula is a powerful tool for computing probabilities because if you know any three of the probabilities in the formula, you can solve for the fourth. It also has some important consequences you are already familiar with.

Earlier in this chapter, you saw the formula $P(E) + P(E^c) = 1$, which relates events and their complements. This is actually just the inclusion-exclusion formula. First, since for any event E, either the event occurs or it doesn't, you know:

$$P(E \cup E^c) = 1$$

Second, because any event and its complement are disjoint, you know:

$$P(E \cap E^c) = 0$$

So, the inclusion-exclusion formula applied to $P(E \cup E^c)$ becomes:

$$\begin{aligned} P(E \cup E^c) &= P(E) + P(E^c) - P(E \cap E^c) \\ 1 &= P(E) + P(E^c) - 0 \\ 1 &= P(E) + P(E^c) \end{aligned}$$

Thus, the formula for computing the complement of an event is just the inclusion-exclusion principle in disguise!

Disjoint Events

Here's another important consequence of disjoint events. Recall the data about favorite colors from Example 5.

Favorite Color	Count
Red	11
Blue	16
Green	7
Yellow	3
Purple	13

Let R be the event that a randomly selected person's favorite color is red, let B be the event that the person's favorite color is blue, and so on. Since a person can only have one favorite color, these events are all disjoint; thus, you can write:

$$P(R \cup B \cup G \cup Y \cup P) = P(R) + P(B) + P(G) + P(Y) + P(P)$$

However, the event $R \cup B \cup G \cup Y \cup P$ represents the entire sample space, so one of these outcomes must occur. This means that $P(R \cup B \cup G \cup Y \cup P) = 1$, and substituting this into the above formula gives you:

$$P(R) + P(B) + P(G) + P(Y) + P(P) = 1$$

You can verify this by computing each individual probability and adding them. From the table, you get that $P(R) = \frac{11}{50}$, $P(B) = \frac{16}{50}$, $P(G) = \frac{7}{50}$, $P(Y) = \frac{3}{50}$, and $P(P) = \frac{13}{50}$, so:

$$\frac{11}{50} + \frac{16}{50} + \frac{7}{50} + \frac{3}{50} + \frac{13}{50} = \frac{50}{50} = 1$$

This is true in general and is another useful property of probability: the sum of all the individual probabilities of each possible outcome must equal 1. One way to think of this is that in an experiment, one of the possible outcomes must occur, so the probability of one of the outcomes occurring is 1, or 100%. This is also related to the fact that the sum of the heights of the bars in a relative frequency histogram must equal 1, or 100%.

Working with the complement and the inclusion-exclusion formula can help make navigating sample spaces and computing probabilities much easier, especially for compound events. As you'll see in the next chapter, compound events play an important role in the study of conditional probability.

Selections With or Without Replacement

When computing probabilities that involve selecting things from a group, it makes a big difference if the selections are made with or without replacement. Here's an example that explains the difference.

Example 8:

From the following data about favorite colors, suppose you selected two individuals and asked both people to name their favorite color. What is the probability that both individuals selected would name blue as their favorite color?

Favorite Color	Count
Red	11
Blue	16
Green	7
Yellow	3
Purple	13

The answer to that question depends on whether you are making selections with or without replacement. Making a selection *with replacement* means that after you make your first choice, you return that item to the set before you select again. Making a selection *without replacement* means that you do not return that item to the set before making your next selection.

In this example, the probability that the first person selected blue as their favorite color is $\frac{16}{50}$. If you are selecting *with replacement*, you return this person's name to the pool of people to choose from, which means you could potentially pick this same person a second time. That would mean that the probability of selecting an individual whose favorite color is blue on your second selection would also be $\frac{16}{50}$.

Because these two events are *independent*—meaning that the outcomes of the choices don't influence each other—you can compute the probability of both events happening by multiplying the probabilities of each individual event. In this case, the probability of choosing an individual whose favorite color is blue twice is:

$$\frac{16}{50} \times \frac{16}{50} = \frac{256}{2{,}500} = 0.1024 = 10.24\%$$

PAINLESS TIP

Events are independent when their results do not influence each other. When events are independent, you can compute the probability of both events happening by multiplying their individual probabilities. You'll learn more about independent events in Chapter 7.

On the other hand, if the selection is made *without replacement*, you cannot pick the first individual a second time because the first individual is not replaced in the pool of people to select from. This means your second choice will be made from a set of 49 individuals. If your first selection is a person whose favorite color is blue, then there will only be 15 individuals left whose favorite color is blue when you make your second selection. This makes the probability of selecting two individuals whose favorite color is blue:

$$\frac{16}{50} \times \frac{15}{49} \approx 0.0980 = 9.80\%$$

Notice that the probability without replacement is slightly lower than the probability with replacement. The removal of one individual whose favorite color is blue reduces the likelihood of selecting an individual whose favorite color is blue a second time.

BRAIN TICKLERS Set #18

1. What is the probability of not rolling a 1 when rolling a fair six-sided die?
2. Suppose you flip a fair coin three times. What is the probability of not seeing any heads?
3. Suppose you choose one of the fifty U.S. states at random. What is the probability that the state's name does not begin with an A?
4. Suppose you roll a pair of fair six-sided dice.

   a. What is the probability that the first die shows a 6?

   b. What is the probability that the second die shows a 6?

   c. What is the probability that at least one die shows a 6?
5. Suppose you roll a pair of fair six-sided dice.

   a. What is the probability the sum of the dice is at least 11?

   b. What is the probability that the sum of the dice is at least 11 and one of the dice shows a 5?

   c. What is the probability that the sum of the dice is at least 11 or one of the dice shows a 5?

   d. What is the probability that the sum of the dice is at least 11 and one of the dice shows a 1?

(Answers are on page 156.)

The Binomial Probability Distribution

The way in which probabilities are distributed over the possible outcomes in the sample space is very similar to the way in which data can be distributed over the possible values it can take. Just as you can imagine a data distribution, you can also imagine a *probability distribution*.

In Chapter 4, you learned about the binomial data distribution, which was introduced by thinking about flipping coins and counting tails. This is also a good way to approach the binomial probability distribution, which plays a crucial role in sampling and inference (topics that will be covered in Chapters 8, 9, and 10).

Computing Binomial Probabilities

Here's an example of computing binomial probabilities, which arise when you perform an experiment repeatedly and count the number of successes or failures.

Example 9:

Suppose you flip a fair coin four times and record the number of tails you see. Elements of the sample space can be thought of as sequences of Hs and Ts that are four letters long: HHHH is flipping four heads, HTHT is flipping heads then tails then heads then tails, and so on. Here's a tree diagram that shows all the possible outcomes.

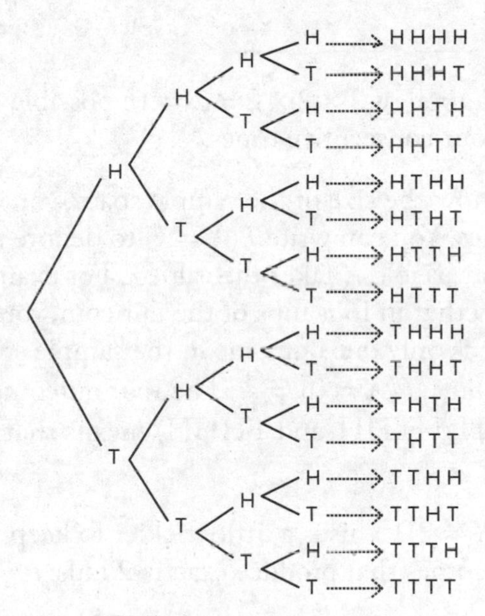

Figure 6–5. Tree Diagram and Sample Space for Four Coin Flips

The number of tails observed can be defined as a *random variable*, a variable that takes values from a random process. If you define the random variable X to be the number of tails observed, then $X = 0$ for the outcome HHHH (zero tails), $X = 2$ for the outcome HTHT (two tails), and so on.

The random variable X can take the values 0, 1, 2, 3, and 4, but these outcomes are not equally likely. Seeing 0 tails is less likely than seeing exactly 1 tail. There is only one outcome in the sample space that produces 0 tails (namely HHHH), whereas there are four outcomes in the sample space that produce exactly 1 tail (namely HHHT, HHTH, HTHH, and THHH).

So, what does the probability distribution of X look like? To compute the probability of each possible value taken by the random variable, first consider the size of the sample space. Each outcome can be represented by a string of Hs and Ts that is four letters long, so the multiplication principle tells you how many possible outcomes there are. When filling in the blanks, you have two choices for the first letter, two choices for the second, two for the third, and two for the fourth.

____ ____ ____ ____

Thus, there are a total of $2 \times 2 \times 2 \times 2 = 16$ possible outcomes of flipping a coin four consecutive times.

Now that you know the size of the sample space, you can compute some probabilities. You can write $P(X = k)$ to denote the probability that the random variable X takes the value k. For example, $P(X = 0)$ is the probability that in four flips of the fair coin, you observe 0 tails. Since there is only one outcome in the sample space (HHHH) that produces 0 tails, $P(X = 0) = \frac{1}{16}$. The four outcomes with 1 tail (HHHT, HHTH, HTHH, and THHH) means that $P(X = 1) = \frac{4}{16} = \frac{1}{4}$.

What about $P(X = 2)$? This is a little trickier to keep track of, but there are six outcomes that produce exactly 2 tails:

HHTT HTHT HTTH THHT THTH TTHH

Therefore, $P(X = 2) = \frac{6}{16} = \frac{3}{8}$.

Finally, determining $P(X = 3)$ and $P(X = 4)$ are easy. Seeing 3 tails is exactly the same as seeing 1 head. There are four such outcomes (HTTT, THTT, TTHT, TTTH), so $P(X = 3) = \frac{4}{16} = \frac{1}{4}$. Seeing 4 tails is the same as seeing 0 heads, and there's only one outcome that produces that (TTTT). Thus, $P(X = 4) = \frac{1}{16}$.

Noticc that $P(X = 0) = P(X = 4)$ and $P(X = 1) = P(X = 3)$. This is an example of the symmetry of the binomial distribution that was discussed in Chapter 4. Also, notice the following:

$$P(X=0)+P(X=1)+P(X=2)+P(X=3)+P(X=4)$$
$$=\frac{1}{16}+\frac{4}{16}+\frac{6}{16}+\frac{4}{16}+\frac{1}{16}=1$$

This makes sense because the sum of all the individual probabilities must equal 1.

Here's a graph of this probability distribution.

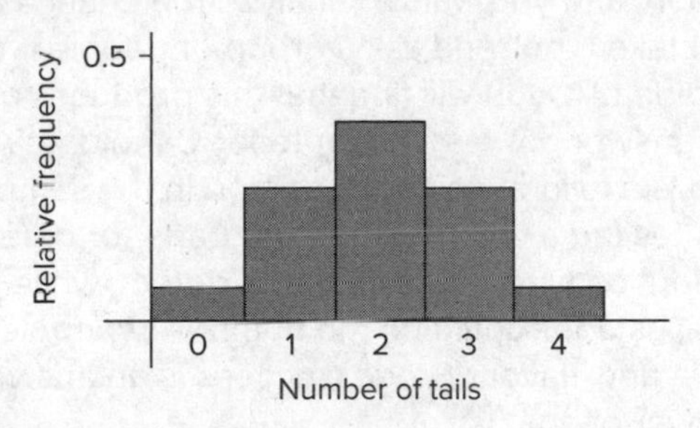

Figure 6–6. Number of Tails Observed in Four Flips

The distribution is symmetric and unimodal, just like the binomial data distribution introduced in Chapter 4. For comparison, here's what the probability distribution looks like for 10 coin flips. (You may recall that this is how the binomial data distribution was introduced in Example 6 of Chapter 4.)

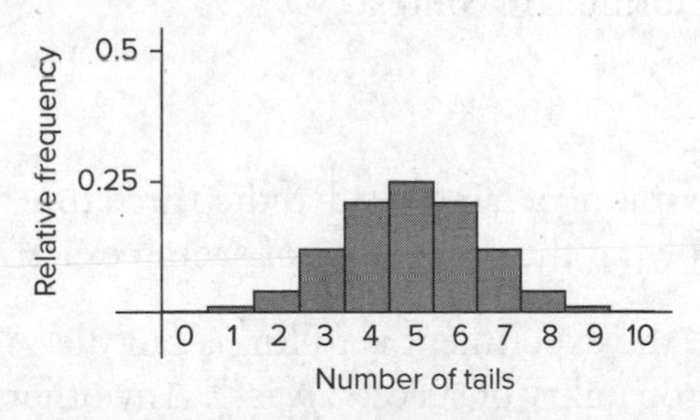

Figure 6–7. Number of Tails Observed in Ten Flips

The binomial probability distribution is the model for binomial data. These distributions arise frequently in probability and statistics. Any time you have an experiment that has two possible outcomes (one that is considered a success and one that is considered a failure), and you perform the experiment repeatedly and independently and count the number of successes, the binomial distribution will appear. In the previous example, the experiment is flipping a coin: a success is flipping tails, and a failure is flipping heads. Each toss of the coin is a *trial*, and your random variable is defined as the number of successes (tails) in each possible outcome.

> There are many real-life examples where the binomial distribution models probability data. For instance, suppose you are interested in an upcoming election, and you want to analyze the chances that Candidate A will win. If you take a poll and ask a group of people if they are going to vote for Candidate A, you will find that this produces a binomial distribution. An answer of "Yes, I will vote for Candidate A" is a success, an answer of no is a failure, and each person in the sample is a trial. In a similar way, testing a collection of auto parts for defects produces a binomial distribution (success is finding a defect, while failure is not finding a defect), as does counting the number of people who would be willing to try a new flavor of soda (success is an answer of yes, while failure is an answer of no).

The Binomial Probability Formula

Binomial distributions are very well understood mathematically, so there are formulas for computing the probability of seeing a given number of successes in a given number of trials. Here's an example to help put that formula in context.

Example 10:

Suppose you are rolling a fair six-sided die three times, and you are interested in knowing the probability of seeing exactly two 6s.

In this situation, the experiment is rolling a fair die. A success is rolling a 6, so the probability of success, p, is $\frac{1}{6}$. Any other roll is a failure, so the probability of failure is $1 - p$, which is $\frac{5}{6}$.

How could you roll exactly two 6s in the three rolls? Well, you could roll a 6, then another 6, and then not a 6. Since each die roll is independent of the others, you can multiply the individual probabilities to compute the probability of this outcome. It would be:

$$\frac{1}{6} \times \frac{1}{6} \times \frac{5}{6} = \left(\frac{1}{6}\right)^2 \left(\frac{5}{6}\right)$$

You could also roll a 6, then not a 6, and then a 6 again. Multiplying these probabilities together gives you:

$$\frac{1}{6} \times \frac{5}{6} \times \frac{1}{6} = \left(\frac{1}{6}\right)^2 \left(\frac{5}{6}\right)$$

Lastly, you could first roll anything other than a 6, then roll a 6, and then roll another 6. The probability of this outcome is $\frac{5}{6} \times \frac{1}{6} \times \frac{1}{6}$, which also equals $\left(\frac{1}{6}\right)^2 \left(\frac{5}{6}\right)$. Since these are all the ways you can get exactly two 6s in three die rolls, to find the total probability, you can add the probabilities of these disjoint events.

$$\left(\frac{1}{6}\right)^2 \left(\frac{5}{6}\right) + \left(\frac{1}{6}\right)^2 \left(\frac{5}{6}\right) + \left(\frac{1}{6}\right)^2 \left(\frac{5}{6}\right) = 3\left(\frac{1}{6}\right)^2 \left(\frac{5}{6}\right)$$

Here, $\frac{1}{6}$ is the probability of success, and it is raised to the power of 2, which is the number of successes you are looking for. The probability of failure, $1 - \frac{1}{6} = \frac{5}{6}$, is raised to the power of 1 because if two of the three trials are successes, then the other one must be a failure. Lastly, the 3 in front represents the number of different ways you could see two successes in a string of three trials:

success, success, failure
success, failure, success
failure, success, success

This process of counting the number of different ways a given number of successes can occur can be generalized to produce the following formula for computing the probability that the random variable X takes the value k:

$$P(X = k) = {}_nC_k \times p^k \times (1 - p)^{n-k}$$

This formula gives the probability of seeing exactly k successes in n trials. It may look intimidating, but this formula just captures what was done in the previous example with the three die rolls. Each variable can be understood as follows: n is the total number of trials, k is the number of successes you are looking for, p is the probability of success, and $1 - p$ is the probability of failure.

The symbol ${}_nC_k$, read "n choose k," is a function that counts the number of ways you can see k successes in n trials. For example, the number of ways you can see exactly two 6s in three rolls of a die is ${}_3C_2 = 3$, and the number of ways you can see exactly two tails in a sequence of four coin flips is ${}_4C_2 = 6$. These ${}_nC_k$ numbers are known as *binomial coefficients*.

The binomial coefficient, ${}_nC_k$, appears often in mathematics. The formula for binomial coefficients is given by ${}_nC_k = \frac{n!}{k!(n-k)!}$, where $n!$ is the *factorial function*. For example, the expression 5! (read "5 factorial") is equal to $5 \times 4 \times 3 \times 2 \times 1 = 120$. The computation for ${}_4C_2$ looks like: ${}_4C_2 = \frac{4!}{2!(4-2)!} = \frac{4!}{2!(2)!} = \frac{4 \times 3 \times 2 \times 1}{2 \times 1 \times (2 \times 1)} = 6$. Calculators with statistics capabilities can compute binomial coefficients for you.

Using the binomial probability formula for $P(X = k)$, you can compute the probability of any number of successes. You can also find the probability of any range of outcomes, like $P(X \geq 3)$ or $P(1 \leq X \leq 5)$. The fact that this formula exists also allows for the derivation of the formulas for mean ($\mu = np$) and standard deviation ($\sigma = \sqrt{np(1-p)}$) of the binomial distribution, which you encountered in Chapter 4.

Another important characteristic of the binomial distribution is that, under the right circumstances, it can be approximated by the normal distribution. When $np \geq 5$ and $n(1 - p) \geq 5$, the normal distribution is a good approximation of the binomial distribution, which makes analyzing binomial data even easier. This fact is related to the important role the binomial distribution plays in the theory of statistical sampling, which you will learn about in Chapter 8.

You can use the binomial distribution when you are counting successes in a series of independent and identical trials where each trial has only two possible outcomes (success or failure) and the probabilities of success and failure remain constant.

BRAIN TICKLERS Set #19

1. Use the properties of the binomial distribution to determine the probability of:

   a. seeing 1 tails in 5 coin flips

   b. seeing no tails in 5 coin flips

   c. seeing more than 1 tail in 5 coin flips

2. Use the properties of the binomial distribution to determine the probability of rolling exactly three 6s in four rolls of a fair die.

3. Suppose an unfair coin comes up tails 90% of the time. Compute the probability of seeing no heads in 10 coin flips.

(Answers are on page 157.)

Brain Ticklers—The Answers

Set #17, pages 137–138

1. $\frac{1}{4}$
2. $\frac{7}{8}$
3. a. $\frac{2}{36} = \frac{1}{18}$

   b. $\frac{4}{36} = \frac{1}{9}$

   c. 7 is the most likely sum, and the probability of a sum of 7 is $\frac{6}{36} = \frac{1}{6}$
4. $\frac{4}{50} = \frac{2}{25}$, or 8% (Alabama, Alaska, Arizona, Arkansas)
5. $\frac{30}{200} = \frac{3}{20}$, or 15%

Set #18, page 148

1. $\frac{5}{6}$
2. $\frac{1}{8}$
3. $1 - \frac{4}{50} = \frac{46}{50} = \frac{92}{100}$, or 92%
4. a. $\frac{1}{6}$

   b. $\frac{1}{6}$

   c. $\frac{11}{36}$
5. a. $\frac{3}{36} = \frac{1}{12}$

   b. $\frac{2}{36} = \frac{1}{18}$

   c. $\frac{3}{36} + \frac{11}{36} - \frac{2}{36} = \frac{12}{36} = \frac{1}{3}$

   d. 0. This is impossible.

Set #19, page 155

1. a. $_5C_1 \times \left(\frac{1}{2}\right)^1 \times \left(\frac{1}{2}\right)^4 = \frac{5}{32} \approx 15.6\%$

   b. $_5C_0 \times \left(\frac{1}{2}\right)^0 \times \left(\frac{1}{2}\right)^5 = \frac{1}{32} \approx 3.1\%$

   c. $1 - {}_5C_0 \times \left(\frac{1}{2}\right)^0 \times \left(\frac{1}{2}\right)^5 = 1 - \frac{1}{32} \approx 96.9\%$

2. $_4C_3 \times \left(\frac{1}{6}\right)^3 \times \left(\frac{5}{6}\right)^1 = \frac{5}{324} \approx 1.5\%$

3. $_{10}C_0 \times \left(\frac{1}{10}\right)^0 \times \left(\frac{9}{10}\right)^{10} \approx 34.9\%$

Chapter 7

Conditional Probability

Beyond the basic properties and formulas of probability is *conditional probability*, which offers a way to compute the probability of an event, given the condition that some other event has occurred. Conditional probability gives you tools to determine how different events influence each other and helps you better analyze data and make inferences. In addition, Bayes' theorem, one of most famous results in probability, can help you reevaluate your beliefs when you encounter new evidence.

The Basics of Conditional Probability

An Example of Computing Conditional Probability

Here's a simple example that demonstrates how knowledge of one event can impact the likelihood of another event and how you can compute the resulting conditional probability.

Example 1:

Suppose you roll a pair of fair six-sided dice and want to know the probability that the sum of those two dice is greater than 9. As calculated in Chapter 6, of the 36 total possible outcomes when you roll two dice, there are six favorable outcomes—$(4, 6)$, $(5, 5)$, $(5, 6)$, $(6, 4)$, $(6, 5)$ and $(6, 6)$—that produce a sum greater than 9. Thus, by the fundamental probability formula, the probability that the sum of the two dice is greater than 9 is $\frac{6}{36} = \frac{1}{6}$.

Now, what if you had some additional information about the rolls? For example, what if you know that the first die roll is a 6? How would that impact the probability that the sum of the two dice is greater than 9? This is the kind of question conditional probability is designed to answer.

If you know that the first roll is a 6, the only missing information is the second roll. If the second die shows a 1, 2, or 3, then the sum of the two dice won't be greater than 9. However, if the second die shows a 4, 5, or 6, then the sum of the two dice will be greater than 9. Thus, given the condition that the first roll is a 6, the sum of the two rolls is greater than 9 in half of the possible outcomes. In this case, you say that the conditional probability that the sum of the two dice is greater than 9, given that the first die roll is 6, is $\frac{1}{2}$.

Notice how the knowledge of the first die roll changed the probability of the event. This makes intuitive sense because if the goal is to get a sum greater than 9, then rolling a 6 on the first die is a very good start. After rolling a 6, all you need is a 4, 5, or 6 on the second die and your sum will be greater than 9.

You can visualize this using the sample space for rolling two fair six-sided dice.

(1,6) (2,6) (3,6) (4,6) (5,6) **(6,6)**

(1,5) (2,5) (3,5) (4,5) (5,5) **(6,5)**

(1,4) (2,4) (3,4) (4,4) (5,4) **(6,4)**

(1,3) (2,3) (3,3) (4,3) (5,3) **(6,3)**

(1,2) (2,2) (3,2) (4,2) (5,2) **(6,2)**

(1,1) (2,1) (3,1) (4,1) (5,1) **(6,1)**

Figure 7–1. Sample Space of a Pair of Six-Sided Dice

Knowing that the first die roll is a 6 restricts your outcomes to the rightmost column. Of the six outcomes in that column, three yield a sum greater than 9 and therefore are favorable. Thus, the probability is $\frac{3}{6} = \frac{1}{2}$. In effect, this additional knowledge changes both the numerator and the denominator in your probability calculation.

On the other hand, if the first die roll was a 1, that would be a really bad start. In fact, getting a sum greater than 9 in that case would be impossible. Since 6 is the highest number you could get on the second roll, there is no way you could get a sum greater than 9. This can also be seen in the sample space.

(1,6) (2,6) (3,6) (4,6) (5,6) (6,6)
(1,5) (2,5) (3,5) (4,5) (5,5) (6,5)
(1,4) (2,4) (3,4) (4,4) (5,4) (6,4)
(1,3) (2,3) (3,3) (4,3) (5,3) (6,3)
(1,2) (2,2) (3,2) (4,2) (5,2) (6,2)
(1,1) (2,1) (3,1) (4,1) (5,1) (6,1)

Figure 7–2. Sample Space of a Pair of Six-Sided Dice

In this case, there are no favorable outcomes in the new restricted sample space. This means that the conditional probability that the sum of the two dice is greater than 9, given that the first die roll is a 1, is 0.

Conditional probability allows you to factor additional information into your determination of the probability of events. You will often encounter conditional probability in the context of coins, dice, and games of chance, but its applications are incredibly wide in scope. For example, how likely is it that a moviegoer will enjoy *Black Widow*, given that they enjoyed *Avengers: Endgame*? How likely is it that a customer will buy a backpack, given that they just bought notebooks and pens? If Tesla's stock price increases, how likely is it that Ford's stock price will also go up? These kinds of questions, asked by researchers, analysts, and scientists every day, are all conditional probability questions.

The Formula for Conditional Probability

Given two events, A and B, the conditional probability of B given A is denoted $P(B|A)$, which is read as "the probability of B given A," or "the conditional probability of B given A." In the case of the experiment in Example 1, if A is the event that the first die roll is a 6 and B is the event that the sum of the two dice is greater than 9, then $P(B|A)$ is the probability that the sum of the two dice is greater than 9, given that the first die roll is a 6. As you saw, $P(B|A) = \frac{1}{2}$ in this experiment. If C is the event that the first die roll is a 1, then $P(B|C) = 0$. That is, the conditional probability that the sum of the two dice is greater than 9, given that the first die roll is a 1, is 0.

For simple experiments, you can sometimes reason out conditional probabilities logically, as in the simple case of the dice above. For the general case, you can use the following formula for conditional probability:

$$P(B|A) = \frac{P(A \text{ and } B)}{P(A)}$$

This formula says that the conditional probability of B given A is equal to the probability of the compound event A and B divided by the probability of the event A. You will also see this formula with $P(A \cap B)$ replacing $P(A \text{ and } B)$:

$$P(B|A) = \frac{P(A \cap B)}{P(A)}$$

PAINLESS TIP

The formula for conditional probability can be rearranged to solve for different probabilities. A common variation is $P(B|A) \times P(A) = P(A \cap B)$. You can also solve for $P(A)$ as $P(A) = \frac{P(A \cap B)}{P(B|A)}$.

Continuing with the experiment from Example 1, the compound event A and B is the event that the first die roll is a 6 and the sum of the two dice is greater than 9, which corresponds to the subset $\{(6, 4), (6, 5), (6, 6)\}$ of the sample space. This subset has size 3, and since the entire sample space has size 36, $P(A \text{ and } B) = \frac{3}{36} = \frac{1}{12}$. The event A is the event that the first die roll is a 6, which corresponds to the subset $\{(6, 1), (6, 2), (6, 3), (6, 4), (6, 5), (6, 6)\}$. This subset has size 6, so $P(A) = \frac{6}{36} = \frac{1}{6}$. Now, you can compute the conditional probability of B given A using the formula:

$$P(B|A) = \frac{P(A \text{ and } B)}{P(A)} = \frac{1/12}{1/6} = \frac{6}{12} = \frac{1}{2}$$

This is the same result you arrived at earlier when reasoning this out logically. In the case of event C, the event that the first die roll is a 1, the conditional probability of B given C is:

$$P(B|C) = \frac{P(B \text{ and } C)}{P(C)}$$

In this scenario, $P(C) = \frac{1}{6}$ since there are 6 outcomes in which the first die roll is a 1. However, $P(B \text{ and } C) = 0$ since there are no outcomes in which the first die is a 1 and the sum of the two dice is greater than 9. Thus, $P(B|C) = \frac{0}{1/6} = 0$, which also confirms the earlier calculation.

REMINDER

An impossible event has probability 0. An event that is guaranteed to occur has probability 1.

To summarize, here are the painless steps for using the formula for conditional probability to compute the probability of one event, given that another event has occurred.

Step 1: Identify the event and the condition. To compute $P(B|A)$, the probability of B given A, B is the event and A is the condition.

Step 2: Compute $P(A \cap B)$, the probability that the event and the condition both happen.

Step 3: Divide the number from Step 2 by $P(A)$, the probability that the condition happens. The result is the conditional probability of B given A.

PAINLESS TIP

The set of all outcomes that occur in both A and B can be written as either $A \cap B$ or $B \cap A$. Because of this, it is also true that $P(A \cap B) = P(B \cap A)$.

Two-Way Tables

Conditional probability forms the basis of a lot of inferential thinking in statistics, as you'll see later in Chapters 9 and 10. It also comes up when working with certain kinds of data sets. A *two-way table* is a way to represent bivariate categorical data, and the analysis of two-way tables relies on the concepts of conditional probability, as seen in the next example.

Example 2:

The following two-way table shows the survey results from high school students who were asked whether they bring lunch from home or buy it at school. What is the probability that a randomly selected 9th grader brings lunch from home?

Grade	Brings Lunch from Home	Buys Lunch at School
9th	80	150
10th	110	130
11th	150	70
12th	160	50

To answer this question, first find the total number of 9th graders by adding the entries in the 9th grade row: $80 + 150 = 230$ total 9th graders. Since 80 of those students bring their lunch from home, the probability that a randomly selected 9th grader brings lunch from home is $\frac{80}{230} = \frac{8}{23}$, or around 34.8%. Notice that this is really a conditional probability question: given the condition that a randomly selected student is a 9th grader, what is the probability that the student brings lunch from home?

As in Example 1, you can visualize the conditional probability by restricting the outcomes. Given that the student is in 9th grade, you only need to consider the 9th-grade row in the table to complete your calculation.

Grade	Brings Lunch from Home	Buys Lunch at School
9th	80	150
10th	110	130
11th	150	70
12th	160	50

Figure 7–3

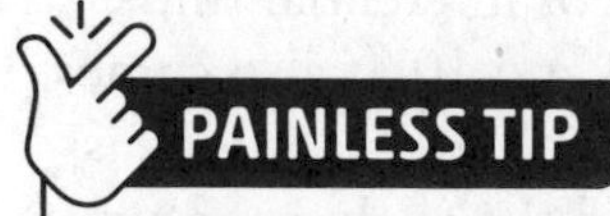

PAINLESS TIP

When solving conditional probability problems that involve two-way tables, it can help to circle the row or column specified by the condition so that you focus only on the data you need for that particular calculation.

You can ask other, more complex, conditional probability questions from a two-way table. For example, in this scenario, what is the conditional probability that a randomly selected student is a 9th grader, given that the student brings lunch from home? The condition that the student brings lunch from home restricts you to the middle column, which represents a total of $80 + 110 + 150 + 160 = 500$ students. Of these 500 students, 80 are 9th graders. So the conditional probability that a randomly selected student is a 9th grader, given that the student brings lunch from home, is $\frac{80}{500} = \frac{4}{25}$, or 16%.

CAUTION—Major Mistake Territory!

When computing probabilities from a two-way table, the favorable and total outcomes you are interested in may span an entire column or row, so don't forget to add them up.

BRAIN TICKLERS Set #20

1. Suppose two fair six-sided dice are rolled.

   a. What is the conditional probability that the sum of the dice is 7, given that the first die roll is a 1?

   b. What is the conditional probability that the sum of the dice is 7, given that one of the rolls is a 1?

   c. What is the conditional probability that the product of the dice is even, given that at least one of the rolls is even?

2. The following two-way table summarizes survey data from a middle school where students from three grade levels were asked how many hours per night they spend doing homework.

Grade	Less Than 1 Hour	1–3 Hours	More Than 3 Hours
6th	170	60	20
7th	120	100	30
8th	55	95	100

   Suppose a student is select at random from the group of students surveyed.

   a. What is the probability that the student is a 6th grader?

b. What is the probability that the student does 1–3 hours of homework per night?

c. What is the conditional probability that the student is an 8th grader, given that the student spends more than 3 hours per night doing homework?

3. Given two tosses of a fair coin, what is the conditional probability that the second flip is tails, given that at least one of the two flips is tails?

(Answers are on page 175.)

Independent Events

In addition to extending the notion of probability and allowing for the analysis of bivariate data through two-way tables, one of the most important properties of conditional probability is that it allows you to understand *independent events*.

The Definition of Independence

Sometimes additional knowledge can increase the probability of an event. As seen in Example 1, the probability that the sum of two fair six-sided dice is greater than 9 is $\frac{1}{6}$, but with the additional information that the first die roll is a 6, this probability increases to $\frac{1}{2}$. You can express this fact using the language of conditional probability. If B is the event that the sum of the two dice is greater than 9 and A is the event that the first die roll is a 6, then $P(B|A) > P(B)$.

You have also seen how additional information can lower the probability of an event. For the event C, that the first die roll is a 1, $P(B|C) < P(B)$. In fact, $P(B|C) = 0$ because it is impossible to get a sum greater than 9 if your first roll is a 1.

An important situation arises when additional knowledge has no impact either way (positively or negatively) on the probability of an event. This happens when the two events are *independent*. As you learned in Chapter 6, informally, this means that the two events are unrelated. So knowing that one event happens gives you no information about whether the other event happens. Technically, independence is a statement about conditional probabilities, as explained in Example 3.

Example 3:

Suppose you are again rolling two fair six-sided dice, as was the case in Example 1, and B is the event that the sum of the two dice is greater than 9. Now, let D be the event that the first die roll is a 4. What is $P(B|D)$?

To compute this, start with the formula for conditional probability:

$$P(B|D) = \frac{P(B \cap D)}{P(D)}$$

The probability that the first die roll is a 4 is just $\frac{1}{6}$, so $P(D) = \frac{1}{6}$. Now, $P(B \cap D)$ is the probability that the first die roll is a 4 and the sum of the two dice is greater than 9. There is only one possible outcome in the sample space that fits these criteria, namely (4, 6), so $P(B \cap D) = \frac{1}{36}$. Plugging these values into the formula for conditional probability gives you:

$$P(B|D) = \frac{1/36}{1/6} = \frac{6}{36} = \frac{1}{6}$$

Notice that $\frac{1}{6}$ is the original probability that the sum of the two dice is greater than 9. That is, $P(B) = \frac{1}{6}$, so $P(B|D) = P(B)$. This means that knowing the first die roll is a 4 has no impact on the probability that the sum of the two dice is greater than 9. The additional information about the first die roll did not change anything. Thus, the events B and D are independent.

This is the technical definition of independence: Two events, A and B, are independent when the conditional probability of B given A is the same as the probability of B. In other words, events A and B are independent when:

$$P(B|A) = P(B)$$

Because of the way conditional probability is defined, if the events A and B are independent, then it is also true that:

$$P(A|B) = P(A)$$

So, when either of these equations is true, both are true, and the events are independent. This means that knowing A occurred has no impact on the probability that B occurs, and vice versa.

When studying basic probability in Chapter 6, you reviewed several examples that involved coin flips. Analyzing these situations becomes even more painless when you establish that fair coin flips are independent of each other.

Example 4:

A person tosses a fair coin twice consecutively. Are those two tosses independent of one another?

Intuitively, you can say that the two tosses of the coin are independent because the outcome of the first toss will have no impact on the outcome of the second toss. Now, you can also confirm this using the formula for conditional probability.

Let T_1 be the event that the first toss comes up tails and T_2 be the event that the second toss comes up tails. To show that these events are independent, start by computing the conditional probability $P(T_2|T_1)$:

$$P(T_2|T_1) = \frac{P(T_2 \text{ and } T_1)}{P(T_1)}$$

The event T_2 and T_1 is seeing tails on two consecutive flips. There is one favorable outcome (TT) out of four possible outcomes (HH, HT, TH, and TT), so this probability is $\frac{1}{4}$. Of course, $P(T_1)$ is just the probability of flipping tails on a fair coin, so $P(T_1) = \frac{1}{2}$. Thus:

$$P(T_2|T_1) = \frac{P(T_2 \text{ and } T_1)}{P(T_1)} = \frac{1/4}{1/2} = \frac{2}{4} = \frac{1}{2}$$

Since $P(T_2)$ is also just the probability of flipping tails on a fair coin flip, $P(T_2) = \frac{1}{2}$, and so $P(T_2|T_1) = P(T_2)$. Knowing that the first coin came up tails does not change the probability that the second coin comes up tails. Therefore, events T_1 and T_2 are indeed independent.

The Multiplication Rule

One of the important consequences of independence has to do with computing probabilities of the intersections of events. If A and B are independent events, then $P(B|A) = P(B)$, and so in the formula for conditional probability:

$$P(B|A) = \frac{P(A \cap B)}{P(A)}$$

you can substitute $P(B)$ for $P(B|A)$ and get:

$$P(B) = \frac{P(A \cap B)}{P(A)}$$

This formula can also be rewritten in the form:

$$P(A \cap B) = P(B) \times P(A)$$

This tells you something very useful about independent events. When two events are independent, the probability of their intersection is the product of their individual probabilities.

PAINLESS TIP

The formula $P(A \cap B) = P(A) \times P(B)$ is known as the *multiplication rule*, but remember that it only applies to independent events.

This fact makes computing certain probabilities very easy. Think about computing $P(T_1 \text{ and } T_2)$, the probability that the first and second coin flips are both tails. In Example 4, this was initially determined by looking at the sample space, but now you can leverage the independence of the events. Since every flip of the coin is independent, T_1 and T_2 are independent events, and so $P(T_1 \text{ and } T_2) = P(T_1) \times P(T_2)$. Thus, $P(T_1 \text{ and } T_2) = \frac{1}{2} \times \frac{1}{2} = \frac{1}{4}$. Similarly, the probability of flipping tails four times in a row would just be $\frac{1}{2} \times \frac{1}{2} \times \frac{1}{2} \times \frac{1}{2} = \left(\frac{1}{2}\right)^4 = \frac{1}{16}$. The independence of the events allows you to compute these probabilities without constructing a sample space and counting favorable outcomes. This principle was used in Chapter 6 to compute binomial probabilities.

CAUTION—Major Mistake Territory!

Many mistakes are made in probability and statistics when people assume that events are independent when they are not. Be careful when you are making this assumption, and be sure that it is justified.

BRAIN TICKLERS Set #21

1. Suppose a fair die is rolled twice.

   a. What is the probability that both rolls are a 1?

   b. What is the probability that both rolls are even?

   c. What is the probability the first roll is a 1 and the second roll is even?

   d. What is the probability the product of both rolls is odd?

2. For each of the following pairs of events, determine whether or not the events are independent.

   a. When rolling two fair six-sided dice, *A* is the event that the first die shows a 3, and *B* is the event that the second die shows a 3.

   b. When rolling two fair six-sided dice, *A* is the event that the first die shows a 6, and *B* is the event that the sum of the rolls is 12.

   c. When flipping four fair coins, *A* is the event that at least three tails are seen, and *B* is the event that four tails are seen.

   d. When drawing a card from a standard deck of playing cards, *A* is the event that the card drawn is a king, and *B* is the event that the card drawn is a heart.

3. Suppose a fair six-sided die is tossed three times.

   a. What is the probability of seeing three 6s in a row?

   b. What is the probability of seeing no 6s in the three rolls?

   c. What is the probability of seeing at least one 6 in the three rolls?

4. Consider the following two-way table.

Grade	Brings Lunch from Home	Buys Lunch at School
9th	80	150
10th	110	130
11th	150	70
12th	160	50

Is bringing your lunch from home independent from being a 9th grader?

(Answers are on pages 175–176.)

Bayes' Theorem

One of the most intriguing results in probability is *Bayes' theorem* (also known as *Bayes' rule*). Bayes' theorem allows you to reverse the direction of conditional probability questions: if you can quantify the impact that A has on B, you can also quantify the impact that B has on A. This allows you to update your prior beliefs when new information is collected.

Before seeing Bayes' theorem in action, it's helpful to first have an understanding of how to compute probabilities by "conditioning" on an event.

Conditioning on an Event

In some situations, events can result from different causes. In these situations, you can compute the probability of your event by considering the various probabilities of the event, given each possible cause. This is known as *conditioning on an event*, and as usual, an example with coin flips is a good way to build an understanding of this concept.

Example 5:

Imagine your friend possesses two identical-looking coins. Coin 1 is a fair coin and shows tails with probability $\frac{1}{2}$. Coin 2, on the other hand, is a weighted coin and show tails with probability $\frac{3}{4}$. Suppose you pick a coin at random from your friend and flip it. What is the probability the coin shows tails?

Let X be the event that you flip tails. To compute $P(X)$, the probability that you flip tails, you have to consider the two possible scenarios: you could choose the fair coin and flip tails, or you could choose the weighted coin and flip tails. These are both conditional probability problems.

Let A be the event that you choose Coin 1. Since the only other option is choosing Coin 2, that event can be denoted A^C.

The complement of an event A is everything else that could happen. It is denoted A^C and read "A complement," or "not A."

If you choose the fair coin, the probability that you flip tails is $P(X|A) = \frac{1}{2}$; this is the conditional probability that you flip tails, given that you select the fair coin. If you choose the unfair coin, the probability that you flip tails is $P(X|A^C) = \frac{3}{4}$; this is the conditional probability that you flip tails, given that you select the weighted coin.

Now, you can compute $P(X)$ by conditioning on the choice of coin. The probability of selecting the fair coin is $P(A)$, so the probability you select the fair coin and flip tails is $P(X|A) \times P(A)$, Similarly, the probabilty you select the unfair coin and flip tails is $P(X|A^C) \times P(A^C)$. Since these two possibilities are disjoint events, the overall probability you flip tails is the sum of these two probabilities:

$$P(X|A) \times P(A) + P(X|A^C) \times P(A^C) =$$

$$\frac{1}{2} \times \frac{1}{2} + \frac{3}{4} \times \frac{1}{2} = \frac{5}{8}$$

So $P(X)$, the probability that you flip tails, is $\frac{5}{8}$. This calculation captures the fact that you could end up flipping tails either by selecting the fair coin and flipping tails or by selecting the unfair coin and flipping tails. The outcome of flipping tails has been *conditioned* on the initial selection of the coin. The fact that $P(X) = \frac{5}{8}$ and is thus greater than $\frac{1}{2}$ makes sense. If you choose the fair coin, you have a 50% chance of flipping tails, and if you choose the unfair coin, your chances are better than 50%. So, overall, your chance of flipping tails with one of the two coins should be better than 50%.

Reversing Conditional Probabilities Using Bayes' Theorem

Bayes' theorem allows you to ask and answer a fascinating question in the kind of situation described in Example 5: If you select a coin at random from your friend and flip tails, what is the probability you chose the fair coin? In other words, what is $P(A|X)$?

Notice the difference between the two conditional probabilities $P(X|A)$ and $P(A|X)$. $P(X|A)$ is the probability that you flip tails, given that you select the fair coin. $P(A|X)$ is the probability that you selected the fair coin given that you ultimately flipped tails. Here, the conditional probability has been reversed, and Bayes' theorem tells you how to compute that reversed conditional probability.

Here is the formula for Bayes' theorem:

$$P(A|B) = \frac{P(B|A) \times P(A)}{P(B)}$$

Notice how $P(A|B)$, the probability of A given B, is expressed in terms of $P(B|A)$, the probability of B given A, as well as the simple probabilities $P(A)$ and $P(B)$. Bayes' theorem lets you trade one conditional probability for another: if you want to know $P(A|B)$, you can compute it using $P(B|A)$.

So, according to Bayes' theorem, $P(A|X)$, the probability that you selected the fair coin given that you ultimately flipped tails, can be computed using this formula:

$$P(A|X) = \frac{P(X|A) \times P(A)}{P(X)}$$

Bayes' theorem allows you to trade a conditional probability you don't know for one you do know. You don't know $P(A|X)$, but you do know that $P(X|A) = \frac{1}{2}$, $P(A) = \frac{1}{2}$, and $P(X) = \frac{5}{8}$, so you can just plug them into the formula to compute $P(A|X)$:

$$P(A|X) = \frac{\frac{1}{2} \times \frac{1}{2}}{\frac{5}{8}} = \frac{\frac{1}{4}}{\frac{5}{8}} = \frac{2}{5}$$

This tells you that, given the fact that you ultimately flipped tails, there is a 40% chance that you chose the fair coin. This may seem odd at first. After all, isn't there a 50% chance of choosing either coin? The answer is yes, but in this conditional probability problem, you have additional information. You know the result is tails, and an outcome of tails is more likely with the weighted coin. This additional knowledge gives greater weight to the possibility that you selected the unfair coin initially, so it makes sense that the chance you chose the fair coin is less than 50%.

This can seem counterintuitive at first, but some extreme case thinking can help resolve the apparent paradox. Suppose instead that the unfair coin always comes up tails. If you choose a coin at random and flip tails, do you think there's a 50% chance you picked the fair coin? What if you flipped the coin 10 times in a row and saw tails every time? Would you still believe the coin you took from your friend was the fair coin? The additional information of seeing tails should change your prior belief about the coin. In this case, the more tails you see, the more convinced you should be that you chose the unfair coin.

In this way, Bayes' theorem gives you mathematical tools for updating your beliefs about situations, such as how likely it is a person is guilty of a crime, given the introduction of a new piece of evidence. You'll also see hints of Bayes' theorem in the study of statistical inference, which is discussed in the next chapter.

In the study of statistics, there are different approaches to thinking about probability. The phrase *Bayesian statistics* refers to an approach where probability represents the strength of your belief about whether or not something will happen. This probability is often referred to as a *prior*, and according to Bayes' theorem, prior probabilities can change in the light of additional information.

BRAIN TICKLERS Set #22

1. Suppose Coin 1 shows tails with probability $\frac{1}{2}$, and Coin 2 shows tails with probability $\frac{3}{4}$. You select a coin at random and flip tails. What is the probability the coin you selected was Coin 2?
2. Suppose Coin 1 shows tails with probability $\frac{1}{2}$, and Coin 2 shows tails with probability $\frac{3}{4}$. You select a coin at random and flip heads. What is the probability the coin you selected was Coin 2?
3. Suppose Coin 1 shows tails with probability $\frac{1}{2}$, and Coin 2 shows tails with probability 1. You select a coin at random and flip heads. What is the probability the coin you selected was Coin 1?

(Answers are on page 176.)

Brain Ticklers—The Answers

Set #20, pages 165–166

1. a. $\frac{1}{6}$

   b. $\frac{2}{11}$

   c. 1. Since even × even = even and even × odd = even, if one of the rolls is even, then the product of the dice must be even.

2. a. $\frac{170 + 60 + 20}{170 + 120 + 55 + 60 + 100 + 95 + 20 + 30 + 100} = \frac{250}{750} = \frac{1}{3}$

   b. $\frac{60 + 100 + 95}{170 + 120 + 55 + 60 + 100 + 95 + 20 + 30 + 100} = \frac{255}{750}$, or 34%

   c. $\frac{100}{20 + 30 + 100} = \frac{100}{150} = \frac{2}{3}$

3. $\frac{2}{3}$. The condition that at least one flip is tails restricts the sample space to {(H, T), (T, H), (T, T)}. In 2 of these 3 outcomes, the second flip is tails.

Set #21, page 170

1. You can use the independence of the two die rolls.

   a. $\frac{1}{6} \times \frac{1}{6} = \frac{1}{36}$

   b. $\frac{1}{2} \times \frac{1}{2} = \frac{1}{4}$

   c. $\frac{1}{6} \times \frac{1}{2} = \frac{1}{12}$

   d. This can only happen if both rolls are odd, so the probability is $\frac{1}{2} \times \frac{1}{2} = \frac{1}{4}$.

2. a. Independent

   b. Not independent

   c. Not independent

   d. Independent

3. Use the independence of events.

a. $\left(\frac{1}{6}\right)^3 = \frac{1}{216} \approx 0.005 = 0.5\%$

b. $\left(\frac{5}{6}\right)^3 = \frac{125}{216} \approx 0.579 = 57.9\%$

c. Use the complement rule with the answer from part (b): $1 - \frac{125}{216} = \frac{91}{216} \approx 42.1\%$

4. No. The probability of bringing your lunch from home, given you are a 9th grader, is $\frac{80}{80 + 150} \approx 35\%$, whereas the overall probability of bringing your lunch from home is $\frac{80 + 110 + 150 + 160}{80 + 110 + 150 + 160 + 150 + 130 + 70 + 50} \approx 55\%$. Since the probabilities are different, the events are not independent.

Set #22, page 174

1. $P(A^C|X) = \frac{P(X|A^C) \times P(A^C)}{P(X)} = \frac{\frac{3}{4} \times \frac{1}{2}}{\frac{5}{8}} = \frac{3}{5}$

2. $P(A^C|X^C) = \frac{P(X^C|A^C) \times P(A^C)}{P(X^C)} = \frac{\frac{1}{4} \times \frac{1}{2}}{\frac{3}{8}} = \frac{1}{3}$

3. 100%. Flipping heads on Coin 2 is impossible.

Chapter 8

Statistical Sampling

In Chapter 5, you learned about the normal distribution and used statistical knowledge and techniques to answer questions such as, "How many adult American men are taller than 6 feet, 1 inch tall?" and "How many test takers scored above 1400 on the SAT?" To answer these questions, you needed statistical knowledge of an entire population, such as the average height of the entire population of adult American men or the average score of the entire population of SAT test takers. In reality, this kind of statistical information—information about an entire population—is usually unknown, and determining it is often the goal when applying statistics. In this chapter, you'll learn how to determine knowledge about a population—such as the mean and standard deviation—through statistical sampling. Sampling is one of the most common statistical practices, and it relies on your knowledge of both data and probability.

Populations and Samples

In the run-up to a major election, it's nearly impossible to avoid seeing polls about who will win. One day, 54% of people say they will vote for Candidate A, and the next day, it's 51% voting for Candidate B. Where do these numbers come from? It isn't possible to ask every person who is voting what their preference is; that would simply be too many people to contact. Instead, a sample of voters are selected and polled.

A *sample* is a subset of a *population*. In an election poll, the population would be the entire collection of people who vote, and the sample is the smaller subset of people who are polled. Asking 1,000 people whom they plan to vote for is much more feasible than asking 100,000,000, which is the point of sampling. Here, 1,000 is the *sample size*. The sample size is usually much less than the size of the population, which makes the sample easier to work with.

What can a sample tell you about the population it comes from? Can knowing the preferences of 1,000 voters really inform you about the preferences of 100,000,000? This is the fundamental question that underlies the theory of sampling, but knowledge of statistics can help you answer it.

Random Samples

In order to apply techniques of statistical sampling to draw appropriate conclusions about a population, you first need your samples to be *representative* of the population. If you want to know how the entire American population will vote in an election, you need to study samples that reflect the entire American population. If you ask 1,000 wealthy people whom they will vote for, their answers might not be representative of the entire population, where only 5% might be considered wealthy. If you ask 1,000 men whom they will vote for, their responses won't be representative of a population that is over 50% women. If your samples are not representative of the population, then the inferences you draw about the population won't be valid.

The simplest way to avoid the problem of unrepresentative samples is to take simple random samples from the population. A *simple random sample* is generated by a random and independent process. In theory, you could imagine having a list of the entire population and then using something like a coin toss or a die roll to choose individuals for your sample. A large enough simple random sample is very likely to be representative of the population, so the conclusions you draw about the population from the sample will be valid.

CAUTION—Major Mistake Territory!

When trying to understand a population by studying samples, make sure the samples are randomly selected. A common mistake is using samples that are not randomly selected. In that case, many of the methods of statistical sampling won't provide an accurate representation of the population.

Once you have a simple random sample, how can you use that sample to draw meaningful inferences about a population? To answer that question, you need to understand parameters and estimators.

Parameters and Estimators

When you're working with a large population, it's generally impossible to know things like the mean and standard deviation. You can't call every single American and ask them how they will vote. You can't measure the height of every adult man and then compute the average. Statistical measures of a population are called *parameters*, and generally speaking, population parameters aren't known.

Population parameters are usually represented by lowercase Greek letters, like μ for the population mean and σ for the population standard deviation. The mean and standard deviation of a sample are usually denoted by $\bar{x}$ and s, respectively.

Often, the goal in statistics is to estimate population parameters. To do this, you study a sample of the population and use statistics from the sample to estimate the corresponding population parameters. For example, you may ask a sample of 1,000 people whom they will vote for to try to estimate whom the entire population will vote for. You might measure a few weeks' worth of commute times to estimate how long an average commute will take in the future. You could measure a stock price every day for a year to try to understand the average stock price overall.

In other words, you use *estimators* to estimate population parameters, and these estimators are computed from samples. Estimators are often referred to as *sample statistics*, or simply, *statistics*. Sometimes, this works just as simply as you might imagine. For example, if you ask 1,000 people whom they will vote for and 60% of the sample say they will vote for Candidate A, then you estimate that 60% of the population will vote for Candidate A. Whether or not this is a good estimate, however, depends on a variety of factors involving the sample, the population, and the estimator itself.

Parameters are quantities related to populations that you can never really know. Use estimators, or statistics, to estimate those population parameters.

Biased and Unbiased Estimators

You also have to consider the estimator itself when determining if an estimate of a population parameter is good or not. An election poll is an example of using a *sample proportion* (the estimator) to estimate the percentage of people in the overall population voting for Candidate A (the population parameter). In another situation, you might try to estimate the mean height of all adult American men (the population parameter) using the mean height of a sample of adult American men (the sample statistic). The sample proportion and the sample mean are common estimators in statistics. Under the right circumstances, they provide very good estimates of their associated population parameter. You'll see examples of these later in this chapter.

However, some estimators that seem natural might not actually provide good estimates. Here's an example of that.

Example 1:

Imagine you want to estimate the highest employee salary in a company. To do this, you take a simple random sample of 10 employees and record their salaries, in tens of thousands of dollars:

41, 38, 55, 120, 71, 46, 39, 61, 98, 55

Employee Salaries (In tens of thousands of dollars)

Given this list, what might you estimate as the highest salary in this company?

A straightforward approach would be to use the maximum of the sample to estimate the maximum of the population. So, you might estimate that the highest salary in the company is $120,000 because that's the highest salary in your sample.

This may seem reasonable, but it isn't going to be a good estimate. Unless you are lucky enough to select the actual highest salary in your random sample, the sample maximum will always underestimate the population maximum. This is an example of a *biased estimator*. If you repeatedly use the maximum of the sample to estimate the maximum of the population, your estimate will be consistently lower than the true value. This consistent error in one direction is bias in your estimate.

In statistical sampling, the most common example of a biased estimator involves standard deviation. You might think that to estimate the standard deviation of a population, you could use the standard deviation of a random sample, but this turns out to be a biased estimator. In general, the spread of a sample will almost always be less than the spread of the population. Because of this, the standard deviation of the sample will almost always underestimate the standard deviation of the population, making it a biased estimator.

Fortunately, it is sometimes possible to counteract the bias in an estimator with a mathematical adjustment. This is exactly what the sample standard deviation, s, is designed to do. Recall the definition of the sample standard deviation from Chapter 3:

$$s = \sqrt{\frac{1}{n-1}\sum(\overline{x} - x)^2}$$

In this formula, you divide by $n - 1$ instead of by N (as you would in the formula for population standard deviation). Dividing by a smaller number produces a larger result; thus, this larger value of s is used to estimate the population standard deviation. This adjusted formula corrects for the bias you would see if you used the formula for population standard deviation.

Sample means and sample proportions are examples of *unbiased estimators*. These sample statistics will produce estimates that, on average, match up with the population parameters that they estimate. You'll see examples of how to use these sample statistics later in this chapter.

BRAIN TICKLERS Set #23

1. Determine whether each of the following are population parameters or sample statistics.

   a. The average salary of a federal employee

   b. The batting average of a baseball player over five games

   c. Your blood pressure as measured during your latest physical exam

2. Suppose you were interested in learning about the life span of electric cars. Give an example of a population parameter you might attempt to estimate.

3. Explain why the following samples might not be representative of the entire population of Americans.

   a. 500 college students

   b. 1,000 subway riders

   c. 100 smartphone users

(Answers are on page 199.)

Sampling Distributions

In the first half of this book, you learned that when you encounter a data set, you should first try to determine the center, spread, and shape of the data. In practice, however, this may be impossible for a large population. You probably cannot collect all the data and calculate the mean and standard deviation for the entire population. You may have no idea what the shape of the data is—it could be normally distributed, uniformly distributed, skewed, or something else entirely. Without the relevant information, how is it possible to develop any useful knowledge of a population? The secret lies in sampling distributions.

When you draw samples from a population and compute sample statistics, you generate data. You can imagine taking many different samples and computing the sample mean for each of them. This set of sample means forms a new data set. It is the *sampling distribution* of the sample mean.

What's remarkable about sampling distributions is that even when you don't know the fundamental statistical facts about a population (such as the mean, the standard deviation, or even the shape of the population distribution), it's possible to recognize these facts about the related sampling distribution. Then, you can use the information about the sampling distribution to estimate the unknown parameters of the population, as seen in the following example.

Example 2:

Suppose you are curious about how much money the average adult American has saved, so you start some statistical analysis. In this situation, the population is all adult Americans, and the variable is "the amount of money in one's savings." For each individual, this variable would take some value, and the set of these values is your population data set. You have no idea what the mean, median, mode, or standard deviation of this data set is. In fact, you don't even know what the shape of the data might be. It might be normal, skewed, or something else entirely.

To try to understand the population, you take a simple random sample. You ask 100 randomly selected adult Americans how much money they have saved. This sample is a data set you can work with. You compute the mean of the sample, that is, the sample mean. Now, imagine taking a second sample of 100 different individuals from the population and computing the sample mean of that set. You would probably get a different number than the first sample mean. If you did this a third, fourth, and fifth time, you would probably get different sample means each time from the different samples. Each of these sample means is a data point in a new data set. You can graph them in a dot plot, as follows.

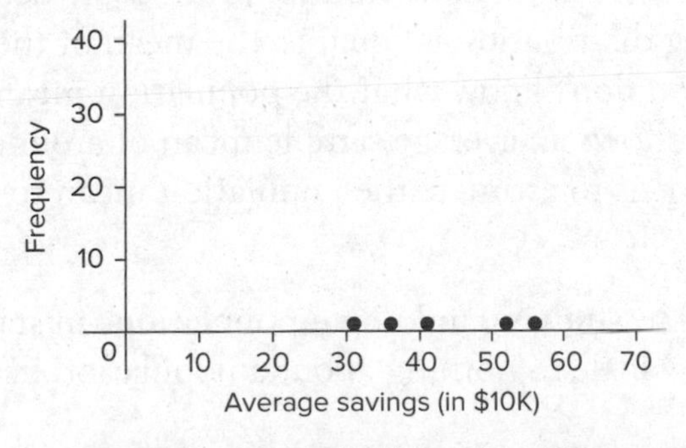

Figure 8–1. Sample Means ($n = 100$)

Now, imagine doing this over and over again, taking hundreds of samples of size 100 and computing and plotting the sample means. Here's a histogram that shows what that data would look like:

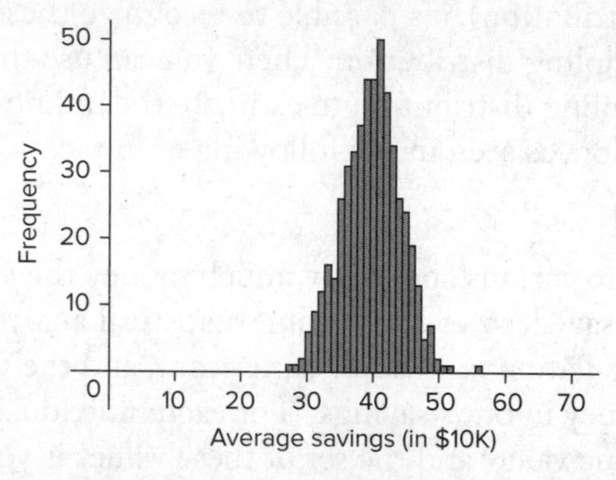

Figure 8–2. Sample Means ($n = 100$)

If you do this for all possible samples of size 100, you would get the sampling distribution of the sample means. This is not the population distribution; this is an entirely new and different data set consisting of sample means from all possible samples of size 100 from the population. Like all distributions, the sampling distribution will be described by its shape and spread. What's fascinating is that statistical theory tells you exactly what to expect from the sampling distribution of the sample means, even though you know virtually nothing about the population the samples were drawn from.

In this scenario involving savings, the sample means will be normally distributed, even though you don't know how the population itself is distributed. You can see this emerging in the histogram above, where the distribution of sample means appears to be symmetrical and unimodal, just like the normal distribution. It is also true that the mean of the sampling distribution is equal to the mean of the population, even though you don't know what the population mean is. Since the histogram shows an average sample mean of around $40,000, you would use this to estimate the population mean to be around $40,000.

This is the key to studying unknown populations in statistics. Even when you know little to nothing about a population itself, you will know something about the relevant sampling distribution. You can then use your knowledge of the sampling distribution to draw conclusions about the population.

Here's another simple example that can make understanding sampling distributions painless.

Example 3:

The probability distribution of a fair six-sided die is uniform because each outcome (1, 2, 3, 4, 5, or 6) is equally likely, as seen in the following figure.

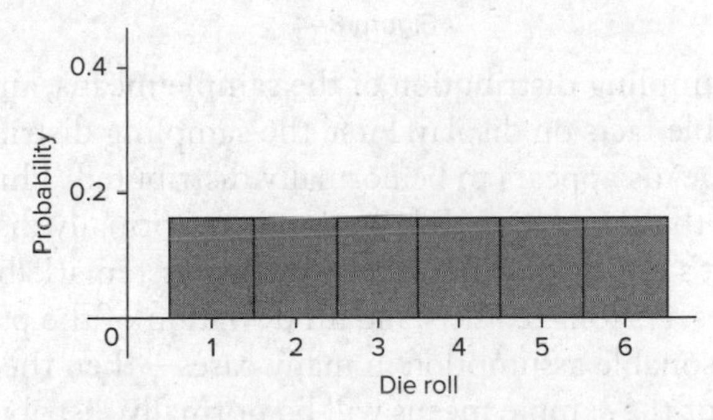

Figure 8–3. Probability Distribution for a Fair Six-Sided Die

You know that the mean of this uniform distribution is just the mean of the outcomes.

$$\frac{1+2+3+4+5+6}{6} = 3.5$$

Now, pretend for a moment that you don't actually know all of this about a fair six-sided die. Pretend that the distribution of the outcomes is the unknown population distribution and that the average die roll is the unknown population mean. How might you estimate the mean? If you had a die, you could roll it a few times and see what the outcomes are. This would be a sample, and the average of those rolls would be the sample mean.

Suppose you roll the die 10 times and compute the average of the rolls. That's taking a sample of size 10 from the population and computing the sample mean. If you do this over and over again and then plot the results in a relative frequency histogram, this is what you'll see:

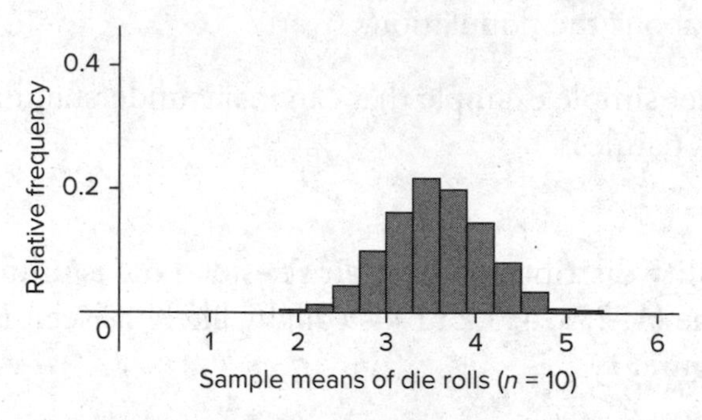

Figure 8–4

This is the sampling distribution of the sample means, and there are two remarkable facts on display. First, the sampling distribution of the sample means appears to be normally distributed. This is true even though the population distribution isn't normally distributed (remember, it's uniform). This fact is true under remarkably general circumstances. As long as the standard deviation of the population is finite—a reasonable assumption in many cases—then the sampling distribution of the sample means will be normally distributed.

Second, notice that the mean of the sampling distribution is 3.5. The variation in the histogram shows you that some samples may have higher means and some may have lower means, but the average sample mean is 3.5. Of course, this is precisely the population mean. In general, the mean of the sampling distribution is equal to the mean of the population because the sample mean is an unbiased estimator of the population mean. Thus, you can estimate the unknown population mean with your sample mean.

These two facts demonstrate the power of statistical sampling. Even if you know virtually nothing about the center, spread, and shape of the population, the distribution of the sample means will be normal. This means you can apply your knowledge about normal distributions to the sampling distribution and obtain knowledge about the population through sampling.

BRAIN TICKLERS Set #24

1. Consider the outcomes of many flips of a fair coin as a population.
   a. Describe the shape of the population.
   b. Suppose samples of 10 tosses of the fair coin are taken, and the number of tails is recorded. Would the distribution of the sample means be uniform?
2. Suppose you want to estimate the age of the oldest living person in the United States, so you select a random sample of 100 Americans. Is the highest age in your sample a good estimate of the age of the oldest living person?
3. Imagine you are playing a game in which there is no limit to how much money you could win or lose. Why would it be difficult to use sampling distributions to understand the possible outcomes of such a game?

(Answers are on page 199.)

Sample Proportions

As you saw earlier in this chapter, unbiased estimators are a powerful tool in studying unknown populations. One particularly useful unbiased estimator is the *sample proportion*. If you are trying to determine what proportion of a population possesses some characteristic, you can use the proportion in a sample of the population—that is, the sample proportion that possesses that characteristic—to estimate it.

1+2=3 MATH TALK!

The word *proportion* has several different meanings in math, but here it means "part of a whole." A proportion in this context will always be a number between 0 and 1, like a probability.

Computing Sample Proportions

Here's an example of how to estimate a population proportion using a sample proportion.

Example 4:

Suppose you are interested in studying public opinion on a new government spending bill. One approach would be to try to understand what percentage, or proportion, of Americans support the spending bill. It's not possible to ask every single American, "Do you support this spending bill?" So this is a situation in which knowledge of the population is impossible. However, it is a perfect opportunity to apply statistical sampling.

You could ask a simple random sample of 100 Americans, "Do you support this spending bill?" Notice that, in this situation, you are collecting categorical data. The response to the question will put each person into a category: supporting the spending bill or not supporting it.

Among this sample, suppose 72 respond yes and 28 respond no. The sample proportion in this case is $\frac{72}{100} = 0.72$, and you would write $\hat{p} = \frac{72}{100}$.

1+2=3 MATH TALK!

The symbol for the sample proportion is $\hat{p}$ and is read "*p* hat" because it looks the *p* is wearing a little hat. The hat symbol is used to indicate estimators in statistics. Without the hat, the symbol would represent the entire population.

The fundamental idea in sampling is to use known sample statistics to estimate unknown population parameters. Here you will use the known sample proportion $\hat{p}$ to estimate the unknown population parameter p. In this case, p is the proportion of all Americans who support the spending bill. This is unknowable, but the sample proportion $\hat{p}$, which equals 0.72, can be used to estimate it. Thus, you would estimate p to be 0.72, that is, you estimate that 72% of Americans support the spending bill.

In summary, to estimate a population proportion, as seen in Example 4, just follow these painless steps.

Step 1: Take a simple random sample of the population.

Step 2: Count the individuals in the sample that match the characteristic you are interested in, and divide that number by the sample size. The result is the sample proportion, $\hat{p}$, and this is your estimate of the population proportion p (the proportion of the population that possesses that characteristic).

Is this a good estimate? The short answer is yes, because the sample proportion is an unbiased estimator of the population proportion. This means there is no systematic underestimating or overestimating associated with the sample proportion. But just how good of an estimate is it? Statistics can help give a clearer answer to this question. Again, the key is using facts about the sampling distribution.

Sampling and Variability

To understand how good your estimate of the population proportion is, you have to understand the effect that variability in sampling can have on the sample proportion. Here's an example that helps illustrate that.

Example 5:

Recall the scenario from Example 4, but now imagine that you took a second sample of 100 American adults and asked them, "Do you support this spending bill?" This time, 63 said yes. In this case, the sample proportion is $\hat{p} = \frac{63}{100} = 0.63$. This is different from the first sample proportion of 0.72, but this isn't cause for alarm. This is a result of the inherent variability in the sampling process.

You have learned that random processes always involve some random variation. The probability of flipping tails on a fair coin is 0.5, but if you flip a coin 100 times, you wouldn't expect to see exactly 50 tails every time. The number will be around 50. If you repeated this experiment over and it over, though, sometimes it will be higher than 50 and sometimes it will be lower.

Taking a random sample of size 100 from a population is just like flipping a coin 100 times. If you take repeated random samples of 100 Americans and ask them if they support this spending bill, you aren't going to get the exact same proportion of yes's every single time. Sometimes, the proportion might be a bit high; other times, it may be a bit low. It's natural for two different samples to produce two different sample statistics because of the variability inherent in the sampling process. The consequence of this variability can be seen in the spread of the sampling distribution, and there are tools for understanding the spread of the data.

If you took repeated samples of 100 Americans, asked them the question, calculated the sample proportion, and plotted the results in a histogram, this is what it would look like:

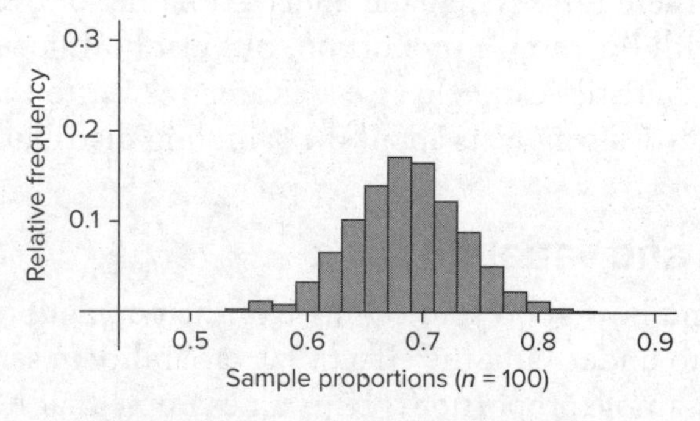

Figure 8–5. Proportion of "Yes" Responses

This is the sampling distribution of sample proportions, and hopefully it looks familiar to you. It is a binomial distribution, and there's a good reason for that. You can think of asking the question "Do you support this spending bill?" as an experiment with two possible outcomes: yes (success) and no (failure). Representing this kind of data is exactly what the binomial distribution is designed for. Instead of flipping a coin 100 times and counting the number of tails, you are asking a question 100 times and counting the numbers of yes's.

This once again demonstrates the power of statistical sampling. Even though you don't know the population parameter p, the true proportion of Americans who support the spending bill, you know that the sample proportion $\hat{p}$ can be used to estimate it. Furthermore, you know that the sampling distribution of the sample proportion $\hat{p}$ is a binomial distribution. Now, you can use all your knowledge of the binomial distribution to analyze and understand the sample proportion $\hat{p}$.

In particular, you can apply your knowledge about the spread of binomially distributed data to compute the standard deviation of the sampling distribution. This is given in terms of the population parameter p and the sample size n:

$$\sqrt{\frac{p(1-p)}{n}}$$

This formula is similar to the formula for the standard deviation of a binomial distribution. It can be used for a sampling distribution as long as the population is much larger than the sample size.

PAINLESS TIP

As a rule of thumb, you can use this formula for the standard deviation of the sampling distribution of a sample proportion when the population is larger than 10 times the sample size.

There is an interesting and important consequence of using this formula. Suppose the true value of the population proportion is $p = 0.68$. Then, for a sample size of $n = 100$, the standard deviation of the sampling distribution would be:

$$\sqrt{\frac{0.68(1-0.68)}{100}} = 0.047$$

Remember, the mean of the sampling distribution is equal to the true population proportion, and the standard deviation tells you how much the sample proportions from the different possible samples of size 100 differ from the mean. In essence, this shows you how accurate the various estimates will be. The closer to the mean, the better the estimate.

Notice what happens if you increase the sample size to $n = 1{,}000$. The standard deviation of the sampling distribution of the sample proportion becomes:

$$\sqrt{\frac{0.68(1-0.68)}{1{,}000}} = 0.015$$

Increasing the sample size decreases the standard deviation of the sampling distribution. When a sample size of 1,000 is used, there is much less variability in your estimates. You can see this when you compare the sampling distributions with different sample sizes.

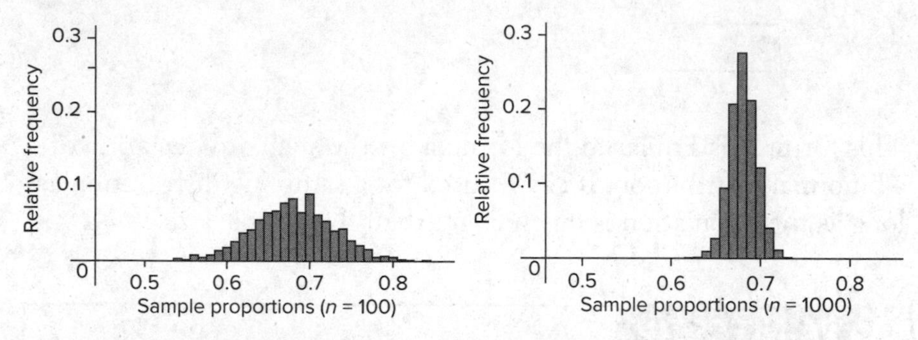

Figure 8–6. Proportion of "Yes" Responses

Figure 8–7. Proportion of "Yes" Responses

Here the two distributions are shown at the same scale. You can see that a larger sample size results in sample proportions that are closer to the mean on average. This is a result of the lower standard deviation. This confirms something that you know intuitively to be true: the more people you ask, the more accurate your poll will be. On average, you will get better estimates using a larger sample size, as you'll see in the next example.

Example 6:

Now, imagine asking only 10 people for their opinion about whether they support the spending bill. Assuming that the true value of the population proportion is $p = 0.68$, you would expect around 7 people to say yes and 3 to say no. However, it wouldn't be unusual for half of your sample to say no. You can compute the chance of this happening using the formula for binomial probability:

$${}_{10}C_5\,(0.68)^5(0.32)^5 = 0.1229$$

REMINDER

The symbol ${}_nC_k$ is read "n choose k" and is equal to $\frac{n!}{k!(n-k)!}$. You can use a calculator to compute this.

A sample where half of the respondents say no would lead to an estimate of the population parameter of 0.50, which is quite far from the true value of 0.68. This is a poor estimate. As the binomial probability calculation above shows, it would happen roughly 12% of the time, or in around 1 out of every 8 samples, even though the vast majority of Americans (68%) support the spending bill.

However, look what happens with a larger sample size. If you take a sample of 100 people instead of 10, the probability that half of them will say no is much, much lower than 12%. Here's the calculation using the formula for binomial probability:

$$_{100}C_{50}\,(0.68)^{50}(0.32)^{50} = 0.000077$$

With this larger sample size, the probability of half your sample saying no is 0.000077, or 0.0077%. This will still lead to a bad estimate of the population parameter, but it won't happen very often; it is about 1,500 times less likely to occur than when the sample size is 10. Furthermore, if you increased the sample size to 1,000, the chances that half would say no when 68% in general would say yes are astronomically low. This is the power of sample size. The larger the sample, the less likely the sample proportion is to deviate much from the true value of the population proportion.

In the case of sample proportions, you know that the sampling distribution is binomial, which allows you to apply your knowledge of the binomial distribution. Under the right circumstances, the binomial distribution can be approximated by the normal distribution, which allows you to then use your knowledge of the normal distribution to study populations and samples. These are facts you will utilize in the next two chapters on confidence intervals and significance testing.

The normal distribution provides a good approximation of the binomial distribution when $np \geq 5$ and $n(1 - p) \geq 5$, where n is the sample size and p is the probability of success. The approximation is better when both quantities are larger than 10.

BRAIN TICKLERS Set #25

1. Repeated samples of 25 flips of a fair coin are taken, and the proportion of tails is recorded for each sample.

   a. What would you expect the mean of the sampling distribution of the sample proportions to be?

   b. Compute the standard deviation of the sampling distribution of the sample proportions.

2. Suppose the true proportion in a population is $p = \frac{2}{3}$. Compute the standard deviation of the sampling distribution for a sample size of:

   a. $n = 10$

   b. $n = 100$

3. Repeated samples of 12 rolls of a fair six-sided die are taken, and the proportion of 6s in each sample are recorded.

   a. What would you expect the mean of the sampling distribution of the sample proportions to be?

   b. Compute the standard deviation of the sampling distribution of the sample proportions.

4. Suppose the true proportion of a population is $p = \frac{1}{2}$. How large a sample would you need to produce a standard deviation of 0.01 in the sampling distribution?

(Answers are on pages 199–200.)

Sample Means

Many of the important facts about sample proportions are also true of sample means, which you used earlier in the section on sampling distributions. Like the sample proportion, the sample mean is an unbiased estimator of the population mean, and the standard deviation of the sampling distribution of the sample means decreases as the sample size increases. Here's a quick example that illustrates how the sample mean relates to the overall population mean.

Example 7:

Suppose you are interested in studying the houschold spending habits of American families, and you decide to focus on grocery bills. How would you determine the average amount of money per month an American family spends on groceries?

As part of your analysis, you might take a random sample from the population of American households and ask how much each family spends on groceries per month. If your sample mean is $\overline{x} = 350$, you would then estimate the population mean to be $\mu = 350$. In other words, your estimate is that the average American family spends \$350 per month on groceries.

How good your estimate is depends on what the sampling distribution looks like. When it comes to sample means, as long as the population standard deviation is finite and the population is much larger than the sample (the rule of thumb is that the population is at least 10 times as large as the sample), then distribution of the sample means will be approximately normal.

The standard deviation of the sampling distribution can be approximated by the following formula (where n is the sample size and σ is the population standard deviation):

$$\frac{\sigma}{\sqrt{n}}$$

This tells you that even if the distribution of monthly grocery bills (from Example 7) is skewed or bimodal, the sampling distribution of the sample means is normal, with a mean equal to the population mean μ and a standard deviation that is approximately $\frac{\sigma}{\sqrt{n}}$.

Just like with sample proportions, you can reduce the variability in your sampling distribution by increasing the sample size. If you take a sample of size 25, the standard deviation of the sampling distribution will be $\frac{\sigma}{\sqrt{25}} = \frac{\sigma}{5}$. However, if you take a sample of size 100, the standard deviation will be $\frac{\sigma}{\sqrt{100}} = \frac{\sigma}{10}$. Remember: A lower standard deviation in the sampling distribution means better estimates of the population parameter.

With sample means, the standard deviation decreases at a rate of around $\sqrt{n}$, where n is the sample size. This means that if you increase the sample size by a factor of 4, you will cut the standard deviation of the sampling distribution in half since $\sqrt{4} = 2$. A smaller standard deviation in the sampling distribution increases the accuracy of the estimator.

The Central Limit Theorem

So much is known about sampling distributions as a result of the central limit theorem, one of the most important results in all of statistics. The *central limit theorem* (CLT) tells you that in many cases, regardless of what the population data is like (uniform, skewed, bimodal, or something else entirely), the sampling distribution of your sample statistic will be approximately normal. This is true as long as the population has a finite mean and a finite standard deviation and as long as the sample size is large enough. (The rule of thumb is that the sample size must be greater than 30 for the central limit theorem to apply.)

Generally speaking, the central limit theorem isn't something you will use, but it is an important result that underlies sampling and inference, so it is good to know about. The central limit theorem tells you that many questions in statistics are really questions about the normal distribution, even if the data you start with isn't normally distributed. This is one of the reasons that the normal distribution is so important in the application of statistics.

The Law of Large Numbers

The central limit theorem is related to another important idea in probability and statistics that you are probably familiar with. You know that if you flip a fair coin, there is a 50% chance of getting tails. This doesn't mean you can predict any one coin flip, but if you flip the coin 100 times, you will expect to see around 50 tails. Even if your first few flips are all heads, you'd still expect around 50 tails overall, because in the long run, you would expect things to even out.

The notion of things evening out in the long run is related to a mathematical theorem known as the *law of large numbers*, which is related to the central limit theorem and sample means. A sample mean can be used to estimate a population mean, but of course, some variation is to be expected. But if you take another sample and average the two sample means, the new average should be closer to the population mean, provided that the samples are independent. If you continue to do this with more and more samples, the average will get closer and closer to the population mean and will equal it in the long run.

Anyone who is interested in sports or games of chance may be familiar with the law of large numbers. For example, a batter in baseball may go through a slump but will likely end up near their career average at the end of the season. A basketball player may miss 10 shots in a row, but over the course of the season, their shooting percentage will rise to their usual career average. A card player may draw unlucky cards several hands in a row, but eventually luck will even out. This is all due to the law of large numbers.

CAUTION—Major Mistake Territory!

Be careful not to misinterpret the law of large numbers. It doesn't tell you that a run of bad luck increases the odds of good luck. Just because a coin comes up heads five times in a row doesn't mean the next flip is more likely to be tails. Each flip is independent, so the chance of tails is still 50%. The law of large numbers never tells you anything is "due" to happen; it is always about long-term behavior.

BRAIN TICKLERS Set #26

1. Suppose a very large population has a population standard deviation of 10. What is the standard deviation of the sampling distribution of the sample means if the sample size is

   a. $n = 25$

   b. $n = 100$

   c. $n = 10{,}000$

2. Given a very large population and a relatively small sample size, by what factor would you have to increase the sample size in order to reduce the standard deviation of the sampling distribution of the sample means by a factor of 10?

3. Which do you think would be a better estimate of the average American's monthly grocery bills: the sample mean of a sample of 100 randomly selected American families, or the sample mean of a sample of 1,000 Americans who shop at a gourmet grocery store?

4. Suppose a person fails his driving test four times in a row. Does the law of large numbers suggest this person will eventually pass this test if he keeps trying?

(Answers are on page 200.)

Brain Ticklers—The Answers

Set #23, page 182

1. a. Population parameter

   b. Sample statistic

   c. Sample statistic

2. You might try to study the number of miles an electric car is driven before it is scrapped or the number of years it is owned before it is replaced.

3. a. College students are typically younger than the average American and thus are not representative of the entire population. Furthermore, of course, not everyone goes to college.

   b. Subways are only present in some large cities, and even in those cities, many people don't ride the subway.

   c. Not everyone owns a smartphone.

Set #24, page 187

1. a. The population distribution is uniform with two equally likely outcomes: heads or tails.

   b. The sampling distribution would not be uniform. Zero tails would be much less likely than 4 or 5 tails.

2. No. The sample maximum is a biased estimator of the population maximum.

3. With the possibility for infinite wins or losses, the population distribution probably won't have a finite standard deviation. If that is the case, then typical sampling techniques cannot be applied.

Set #25, page 194

1. a. $\frac{1}{2} = 0.5$

   b. $\sqrt{\frac{0.5 \times 0.5}{25}} = 0.1$

2. a. $\sqrt{\dfrac{\frac{2}{3}\times\frac{1}{3}}{10}} \approx 0.15$

   b. $\sqrt{\dfrac{\frac{2}{3}\times\frac{1}{3}}{100}} \approx 0.05$

3. a. $\dfrac{1}{6}$

   b. $\sqrt{\dfrac{\frac{1}{6}\times\frac{5}{6}}{12}} \approx 0.11$

4. Solve the equation $\sqrt{\dfrac{\frac{1}{2}\times\frac{1}{2}}{n}} = 0.01$ to find that $n = 2{,}500$.

Set #26, page 198

1. a. 2

   b. 1

   c. 0.1

2. You would need to increase the sample size by a factor of 100 to decrease the standard deviation by a factor of 10 since $\sqrt{100} = 10$.
3. The smaller, random sample will provide a better estimate than the larger, biased sample.
4. No. The individual driving tests are not independent events, so the law of large numbers does not apply. The individual results are related by some underlying condition, like lack of preparation or knowledge.

Chapter 9

Confidence Intervals

In the last chapter, you learned how to apply statistics to analyze unknown populations. By studying random samples from a population, you can use the known properties of sampling distributions to estimate unknown population parameters. But how do you know whether those estimates are good? As you'll learn in this chapter, statistics can answer that question, too.

Making Inferences from Data

Statistical sampling allows you to use sample statistics (like the sample mean and the sample proportion) to estimate population parameters (like the population mean and population proportion). This is an example of using statistics to draw inferences from data. The population parameters may be theoretically or practically unknowable, but statistical techniques can help you infer some of their characteristics. The following example shows you how to make simple inferences from data.

Example 1:

Suppose you are interested in how much sleep American high school students get each night. You can imagine asking every high school student in the country, "How many hours of sleep did you get last night?" which would produce a quantitative data set. In theory, this data set has a center, a spread, and a shape that you could try to determine. In practice, though, you can't ask every single high school student in America because there are too many individuals in the population. Instead, you could pose this question to a random sample of 100 American high school students and use their responses to form a sample data set.

Once you have your sample data, you can compute the sample mean. This is the average hours of sleep the students from the sample got last night. Suppose the sample mean is 7.5 hours. Since the sample mean is an unbiased estimator of the population mean, you estimate the population mean to be 7.5 hours. But how good is this estimate?

As you learned in Chapter 8, sampling naturally involves some random variation. Different random samples of 100 American high school students will produce different sample means. One sample of 100 students might have a mean of 7.2 hours, another might have a mean of 6.7 hours, a third may have a mean of 8.1 hours, and so on. Each of these different sample means would lead to a different estimate of the population parameter. The natural variation inherent in sampling indicates a level of uncertainty about the process. An important part of using statistics to draw inferences is quantifying this kind of uncertainty. One way to do that is to attach a *margin of error* to your estimate.

For example, a sample mean of 7.5 hours leads to an estimate of 7.5 hours for the population mean, but maybe your knowledge of the variability of the sampling distribution suggests a margin of error of 0.4 hours. This means that your best estimate of how much sleep a high school student gets on average is 7.5 hours, but you wouldn't be surprised if the average actually turned out to be anywhere from 7.1 hours to 7.9 hours.

This new kind of estimate—given with a margin of error—is really an interval of estimates. It is more honest, and more useful, than providing a single number because it acknowledges the uncertainty of your estimate and attempts to quantify it. As you'll learn in the next section, it is common to report statistical estimates as intervals. Like everything else in statistical sampling, computing these intervals depends on understanding sampling distributions.

Estimating with Confidence

The estimate that the average amount of sleep an American high school student gets per night is between 7.1 hours and 7.9 hours (as calculated in Example 1) is an example of a confidence interval. In contrast to a *point estimate*, which is a single number, a *confidence interval* provides a range of values where the population parameter might lie and attaches a level of confidence to that estimate.

Many familiar words have specific, technical meanings in mathematical contexts that differ from their everyday usage. This is especially true in statistics, where common words like "normal" and "average" have very specific meanings. "Confidence" is another such word. In statistics, the meaning of confidence is aligned with its usual meaning—a measure of how certain you are of something—but the certainty is about procedure rather than about an outcome. As you'll see later in this chapter, statistical confidence refers to how frequently you expect a particular procedure to produce a correct answer.

Defining Confidence Intervals

Confidence intervals are one of the most common ways to use statistics to make an estimate or a projection. Presenting an estimate as an interval acknowledges the uncertainty in the process of estimating and attempts to quantify it by providing a range of values instead of just a single value. Determining how large that range should be depends on the margin of error of your estimate, and that, in turn, depends on properties of sampling distributions.

Recall the following key facts about sampling distributions from Chapter 8:

- Under the right circumstances, the sampling distribution of a sample statistic can be approximated by a normal distribution, even if the distribution of the population is unknown.
- The standard deviation of the sampling distribution is related to the standard deviation of the population, and it gets smaller as n, the sample size, gets larger.

These facts can be used together to compute an appropriate margin of error for your estimate and to provide a level of confidence about the estimated location of the population parameter. When you are trying to estimate an unknown population mean, μ, these facts give you a lot of information to work with. You know that the sampling distribution of the sample means is normal, has a mean equal to the population mean, μ, and has a standard deviation of $\frac{\sigma}{\sqrt{n}}$ (where σ is the population standard deviation and n is the sample size). Knowing about the center, spread, and shape of the sampling distribution allows you to determine just how accurate your estimate is.

Generally speaking, you probably won't know the population standard deviation, but sometimes additional knowledge allows you to estimate it, which will help you make inferences. For example, a number of scientifically conducted sleep studies may have determined that a reasonable estimate for the standard deviation (of the number of hours of sleep per night an American high school student gets) is around 2 hours. You can use this as a stand-in for the unknown population standard deviation, which together with the known sample size of $n = 100$ allows you to compute the standard deviation of the sampling distribution:

$$\frac{\sigma}{\sqrt{n}} = \frac{2}{\sqrt{100}} = \frac{2}{10} = 0.2$$

So, now you know that your sampling distribution is normal, has mean μ, and has a standard deviation of 0.2.

The original sample of 100 high school students had a sample mean of 7.5 hours. The relevant question is this: where does 7.5 hours fall in the sampling distribution? If this is close to the mean of the distribution, then this will be a good estimate of the population parameter. If it's far away from the mean, then it is a poor estimate.

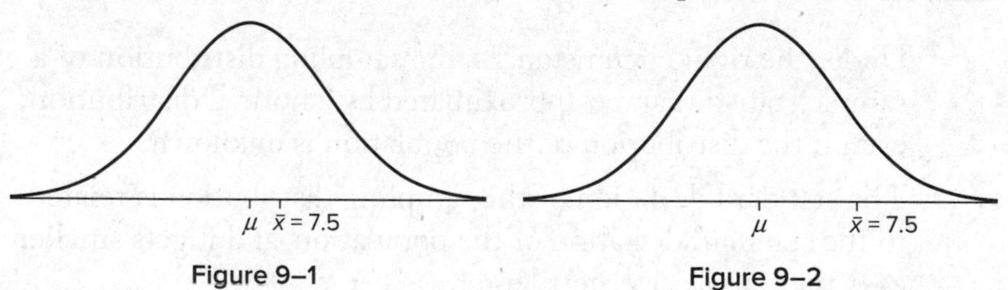

Figure 9–1   Figure 9–2

Of course, you can't say for sure. However, you can use knowledge of the normal distribution to make sense of the situation.

Constructing Confidence Intervals

Recall that, by the empirical rule, 68% of the data in a normal distribution lies within one standard deviation of the mean. In the context of the sampling distribution from Example 1, this tells you that in 68% of the possible samples of 100 high school students, the sample mean is within one standard deviation of the mean of the distribution. And you know the mean of the sampling distribution: it's the population mean, μ, because the sample mean is an unbiased estimator.

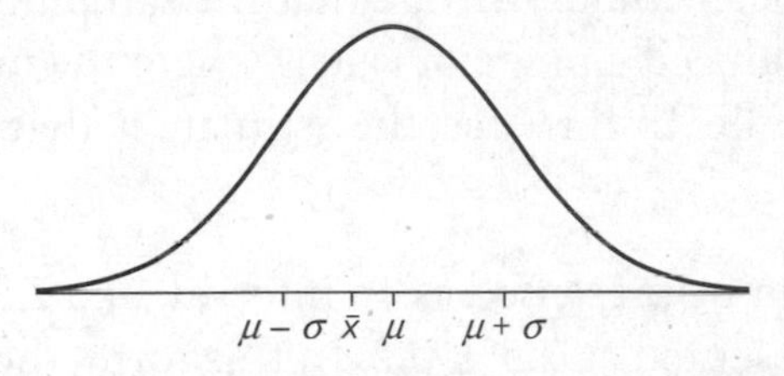

Figure 9–3

Since the standard deviation of this particular sampling distribution is 0.2, then for 68% of the possible samples you could take, the sample mean, $\bar{x}$, lies in the interval from $\mu - 0.2$ to $\mu + 0.2$. This can be written as an inequality:

$$\mu - 0.2 \leq \bar{x} \leq \mu + 0.2$$

If your sample mean of $\bar{x} = 7.5$ happens to be one of the 68% of the samples that has this property, this inequality becomes:

$$\mu - 0.2 \leq 7.5 \leq \mu + 0.2$$

This tells you that 7.5 is close to μ: 7.5 can't be bigger than $\mu + 0.2$, and it can't be less than $\mu - 0.2$. But if 7.5 is close to μ, then μ is close to 7.5. How close? Some simple math can cleverly turn these inequalities around.

The compound inequality $\mu - 0.2 \leq 7.5 \leq \mu + 0.2$ is really these two inequalities:

$$\mu - 0.2 \leq 7.5 \text{ and } 7.5 \leq \mu + 0.2$$

Rearranging these inequalities gives you:

$$\mu \leq 7.7 \text{ and } 7.3 \leq \mu$$

Putting them back together gives you a single compound inequality that tells you where μ should be:

$$7.3 \leq \mu \leq 7.7$$

In short, if 7.5 is within 0.2 of μ, then μ is within 0.2 of 7.5. This places μ between $7.5 - 0.2 = 7.3$ and $7.5 + 0.2 = 7.7$.

This is a confidence interval. The known sample mean and standard deviation, together with knowledge that the sampling distribution is normal, have produced a range of values where the unknown population mean might lie. In this case, the estimate is that μ is somewhere between 7.3 and 7.7.

This estimate can be presented as an interval, like $7.3 \leq \mu \leq 7.7$, but it can also be presented as 7.5 ± 0.2. In this form, the center of the confidence interval, 7.5, is highlighted, and the 0.2 is interpreted as the *margin of error* of the estimate. According to your estimate, the population mean is around 7.5, but it could be as high as 7.7 or as low as 7.3.

If you've ever seen a political poll, you've heard the phrase "margin of error." This indicates that the results of the poll are being presented as a confidence interval.

Understanding Confidence

The "confidence" part of "confidence interval" is subtle. Remember, the original inequality, $\mu - 0.2 \leq \overline{x} \leq \mu + 0.2$, is true for 68% of the possible samples of size 100 that you could draw from the population. If your sample, the one with the sample mean of 7.5, happens to be among that 68%, then the inequality $7.3 \leq \mu \leq 7.7$ is true. In this case, you have correctly identified a range for the population mean.

On the other hand, if by random chance your sample isn't among that 68%, then this interval will not contain the population mean, and your estimate will not be correct. The uncertainty of your estimate is really the uncertainty about this particular sample: is it one of the 68% within one standard deviation of the mean, or is it one of the other 32% that isn't? Assigning a *confidence level* to your estimate is a way to quantity this uncertainty. Here, there is a 68% level of confidence because if your sample was among 68% of all possible samples, then it leads to a correct estimate.

This notion of confidence level can easily be misunderstood. A 68% confidence level does not mean that there is a 68% chance that the population mean is between 7.3 hours and 7.7 hours. Confidence levels are frequently misinterpreted in this way, but such statements don't really make sense. You may not know what the population mean is, but it's still some fixed number. Saying there's a 68% chance that a fixed number is between 7.3 and 7.7 doesn't make sense. The number is either between 7.3 and 7.7 or it isn't.

The 68% isn't about the population parameter; it's about the possible samples. A confidence level is a claim about the sampling distribution. If you repeatedly took random samples of 100 high school students, computed the sample mean, and constructed confidence intervals in this way, the population mean would lie in 68% of those confidence intervals. This is often demonstrated with this figure.

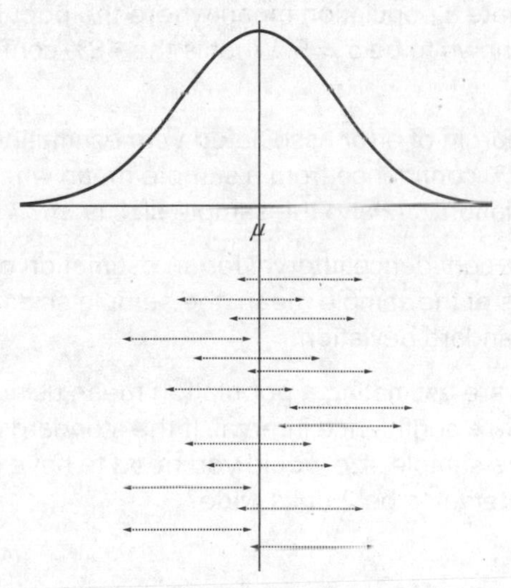

Figure 9–4

This image shows the normal sampling distribution on top and 15 confidence intervals computed from 15 different random samples drawn from the population. Among the 15 confidence intervals shown, 10 of them (around 67%) contain the population mean, and 5 of them don't. The confidence level tells you how often this process successfully works to correctly pin down the location of the population mean.

CAUTION—Major Mistake Territory!

When reporting confidence intervals, don't say, "There is a 68% chance the population mean lies in this interval." Say instead, "I am 68% confident that the unknown population mean lies in this interval." The confidence level is about the process, not the location of the parameter.

Now, 68% might not seem like a high level of confidence, but as a result of how easy it is to work with the normal distribution, you can construct confidence intervals for any level of confidence. You'll learn how in the next section.

BRAIN TICKLERS Set #27

1. Suppose a sample mean of $\bar{x} = 50$ from a sample of size 100 is used to estimate a population mean where the population standard deviation is known to be $\sigma = 5$. What is the 68% confidence interval for this estimate?
2. What is the margin of error associated with estimating a population mean with 68% confidence from a sample mean where the population standard deviation is 12 and the sample size is 25?
3. Write the 68% confidence interval for an estimation of a population mean in terms of the sample mean, the sample size, and the population standard deviation.
4. Suppose you are estimating a population mean using the sample mean and a 68% confidence interval. If the standard deviation is 10, how large a sample size would you need to have for the 68% confidence interval to be 2 units wide?

(Answers are on page 220.)

Different Levels of Confidence

In Example 1, you constructed a 68% confidence interval to estimate a population mean, but generally speaking, the most commonly used confidence level is 95%. This is because 95% is a high level of confidence and is also a convenient number when it comes to the normal distribution.

95% Confidence Intervals

Recall that, by the empirical rule, 95% of the data in a normal distribution lies within two standard deviations of the mean. This makes 95% confidence intervals easy to construct: you just add and subtract two sample standard deviations from your estimate.

To construct a 95% confidence interval for a population mean, just follow these painless steps.

Step 1: Compute the sample mean, $\overline{x}$, of your simple random sample.

Step 2: Compute the standard deviation of the sample, which is $\frac{\sigma}{\sqrt{n}}$.

Step 3: Subtract $2 \times \frac{\sigma}{\sqrt{n}}$ from the sample mean $\overline{x}$. This is the lower bound of the confidence interval.

Step 4: Add $2 \times \frac{\sigma}{\sqrt{n}}$ to the sample mean $\overline{x}$. This is the upper bound of the confidence interval.

A 95% confidence interval can therefore be represented as $\overline{x} \pm 2\frac{\sigma}{\sqrt{n}}$. Notice that this means $2\frac{\sigma}{\sqrt{n}}$ is the margin of error of a 95% confidence interval.

PAINLESS TIP

When estimating a population mean, the margin of error is always a multiple of $\frac{\sigma}{\sqrt{n}}$, which means you reduce the margin of error of your estimate by increasing the sample size, *n*. This is related to how increasing the sample size reduces variability in the sampling distribution.

Here's an example of the painless steps for constructing a 95% confidence interval in action.

Example 2:

Recall the data from Example 1, where you were interested in how many hours of sleep American high school students get each night. From that data, you concluded that $\overline{x} = 7.5$ and $\frac{\sigma}{\sqrt{n}} = \frac{2}{\sqrt{100}} = 0.2$. To construct a 95% confidence interval for this data, just follow the painless steps.

You've already computed the sample mean and the standard deviation, so Steps 1 and 2 are done. Now, subtract 2×0.2 from 7.5 to get the lower bound of your interval, and add 2×0.2 to 7.5 to get the upper bound:

$$7.5 - 2 \times 0.2 \leq \mu \leq 7.5 + 2 \times 0.2$$
$$7.5 - 0.4 \leq \mu \leq 7.5 + 0.4$$
$$7.1 \leq \mu \leq 7.9$$

This 95% confidence interval means that this procedure will produce an interval that contains the population mean 95% of the time. You can say that there is a 95% confidence level associated with the claim that the unknown population parameter μ satisfies the inequality $7.1 \leq \mu \leq 7.9$.

Notice that the 95% confidence interval, which ranges from 7.1 to 7.9, is larger than the 68% confidence interval, which ranges from 7.3 to 7.7. This should make sense: in order for the process to work correctly for more of the possible samples, you have to account for more potential variability. In other words, if you want to be more confident about your estimate, you need a bigger margin of error.

You'll see this kind of tradeoff a lot in statistical reasoning. To reduce uncertainty, you need to account for more possibilities. An extreme example helps illustrate this. You can be 100% confident that the average amount of sleep an American high school student gets is between 0 and 24 hours. A more useful estimate will narrow that range down, but that increases the chance that your estimate won't be correct: the smaller your window of prediction, the less certain you can be about it. Every useful estimate involves uncertainty, but the important feature of a confidence interval is that you acknowledge the uncertainty and try to quantify it.

Level C Confidence Intervals

You've seen examples of 68% confidence intervals and 95% confidence intervals, but thanks to the known properties of the normal distribution, you can construct a confidence interval for any desired level of confidence. A *level* C *confidence interval* is constructed so that the population parameter will lie in that interval with probability C. To construct a level C confidence interval, the strategy is simple: use the interval in the standard normal distribution that contains $C\%$ of the data, and convert the corresponding z-scores into data values in your sampling distribution. Here's an example that shows you how this is done.

Example 3:

Suppose you are studying how much time adults spend on their phones. You take a sample of 100 adults who use smartphones and record the number of times they check their phone in a day. The sample mean of your data is 130, and you want to turn this into a 90% confidence interval estimate of the population mean.

To do this, you need to know the population standard deviation, so imagine that previous research suggests that the standard deviation for this kind of data is around 60. You can now approximate the sampling distribution of the sample mean by a normal distribution with mean μ and standard deviation $\frac{\sigma}{\sqrt{n}} = \frac{60}{\sqrt{100}} = 6$. To construct a 90% confidence interval, however, you must first turn to the standard normal distribution.

The standard normal distribution is the normal distribution with a mean of $\mu = 0$ and a standard deviation of $\sigma = 1$.

Your goal is to find an interval centered at the mean that contains 90% of the area under the standard normal curve. Since the interval must be symmetric about $\mu = 0$, you need 45% of the data to lie on either side.

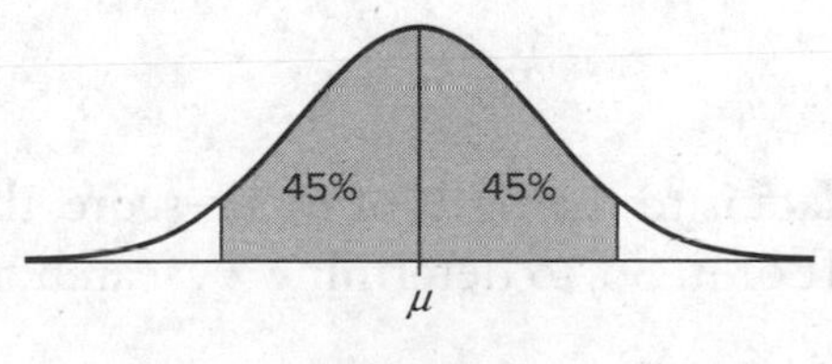

Figure 9–5

All you need are the z-scores at either end of this interval, and the standard normal table can help you find them.

REMINDER

The entries in the standard normal table tell you the percentage of the data that lies to the left of a particular z-score. For example, in the standard normal table, a z-score of 1 corresponds to 0.8413, which tells you that 84.13% of the data lies to the left of $z = 1$.

Since the normal distribution is symmetric, 50% of the data lies above the mean. So if you want an interval that covers 45% of the data above the mean, that leaves 5% of the data to the right of that z-score. This z-score is known as the *upper 5% critical value*, or just the *upper critical value*, and it is sometimes denoted z^*.

PAINLESS TIP

The upper critical value for a level C confidence interval is the z-score where $\frac{1-C}{2}$ of the data under the standard normal curve lies to the right of that z-score. In this formula, C must be interpreted as a number between 0 and 1, so for a 90% confidence interval, C would be 0.90.

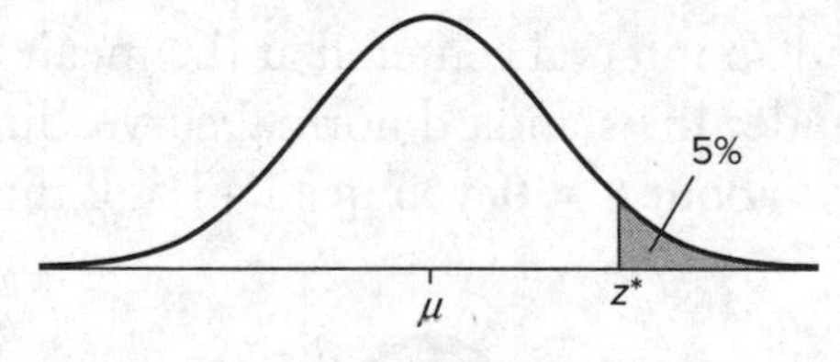

Figure 9–6

Since 5% of the data is to the right of this z-score, then 95% of the data lies to the left of it. So, to determine z^*, search the standard normal table for 0.95.

z	0.00	0.01	0.02	0.03	0.04	0.05	0.06	0.07	0.08	0.09
1.2	0.8849	0.8869	0.8888	0.8907	0.8925	0.8944	0.8962	0.8980	0.8997	0.9015
1.3	0.9032	0.9049	0.9066	0.9082	0.9099	0.9115	0.9131	0.9147	0.9162	0.9177
1.4	0.9192	0.9207	0.9222	0.9236	0.9251	0.9265	0.9279	0.9292	0.9306	0.9319
1.5	0.9332	0.9345	0.9357	0.9370	0.9382	0.9394	0.9406	0.9418	0.9429	0.9441
1.6	0.9452	0.9463	0.9474	0.9484	0.9495	0.9505	0.9515	0.9525	0.9535	0.9545
1.7	0.9554	0.9564	0.9573	0.9582	0.9591	0.9599	0.9608	0.9616	0.9625	0.9633
1.8	0.9641	0.9649	0.9656	0.9664	0.9671	0.9678	0.9686	0.9693	0.9699	0.9706
1.9	0.9713	0.9719	0.9726	0.9732	0.9738	0.9744	0.9750	0.9756	0.9761	0.9767
2.0	0.9772	0.9778	0.9783	0.9788	0.9793	0.9798	0.9803	0.9808	0.9812	0.9817
2.1	0.9821	0.9826	0.9830	0.9834	0.9838	0.9842	0.9846	0.9850	0.9854	0.9857

Figure 9–7

There may be no exact entry of 0.95 in the standard normal table, so take the best approximation you can find. For example, the entry for a z-score of 1.65 is 0.9505. This means that roughly 95% of the data lies to the left of $z = 1.65$, so roughly 5% of the data lies to the right of $z = 1.65$. This makes the upper critical value $z^* = 1.65$. By the symmetry of the normal distribution, around 5% of the data should lie below $z = -1.65$. You have now found the interval you were looking for: in the standard normal table, 90% of the data lies between $z = -1.65$ and $z = 1.65$.

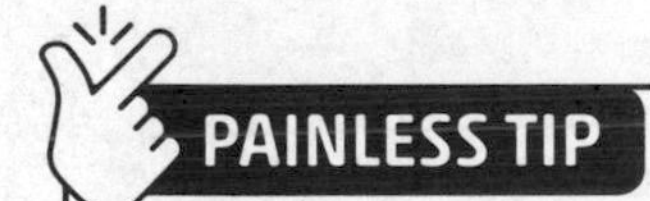

Calculators with statistical capabilities can directly compute z-scores associated with specific probabilities using “inverse” normal CDF functions (where CDF stands for cumulative distribution function, which you learned about in Chapter 2). This can replace the need to look up the value in the standard normal table.

Thus, to contain 90% of the data under the standard normal curve, an interval must extend 1.65 standard deviations above and below the mean. The same will be true in the normal sampling distribution as well, so you can work backward to construct your 90% confidence interval.

Your initial estimate of the population mean is the sample mean, which is 130. Your analysis of the standard normal table gives you the margin of error of 1.65 standard deviations. Since the population standard deviation was assumed to be 60, the standard deviation of the sampling distribution is $\frac{\sigma}{\sqrt{n}} = \frac{60}{\sqrt{100}} = 6$, which makes the 90% confidence interval:

$$130 \pm 1.65 \times 6$$

Since $1.65 \times 6 = 9.9$, you can round the margin of error to 10 and get a confidence interval of 130 ± 10. When expressed as an inequality, this is:

$$130 - 1.65 \times 6 \leq \mu \leq 130 + 1.65 \times 6$$

$$130 - 10 \leq \mu \leq 130 + 10$$

$$120 \leq \mu \leq 140$$

This means that in 90% of the samples of 100 adults, the mean number of times the adult checks their phone will be between 120 and 140. You can say that you are 90% confident that the population mean is between 120 and 140.

To recap this example, here are the painless steps for constructing a level C confidence interval for a population mean using a sample mean $\overline{x}$ from a sample of size n from a population with standard deviation σ.

Step 1: Compute $\frac{(1-C)}{2}$, making sure to interpret C as a number between 0 and 1 (for example, 0.90 instead of 90%).

Step 2: Subtract $\frac{(1-C)}{2}$ from 1, and search for this entry in the standard normal table. If there is no entry that matches exactly, take the best approximation you can find. The z-score that aligns with this entry is z^*, the upper critical value.

Step 3: Compute the margin of error, $z^*\frac{\sigma}{\sqrt{n}}$.

Step 4: Add and subtract the margin of error (from Step 3) from the sample mean $\overline{x}$ to determine the confidence interval: $\overline{x} \pm z^*\frac{\sigma}{\sqrt{n}}$.

BRAIN TICKLERS Set #28

1. Construct a 95% confidence interval to estimate a population mean. given a sample mean of $\overline{x} = 100$, a population standard deviation of $\sigma = 20$, and a sample size of $n = 100$.
2. Construct a 99% confidence interval to estimate a population mean, given a sample mean of $\overline{x} = 100$, a population standard deviation of $\sigma = 20$, and a sample size of $n = 100$.
3. Construct a 95% confidence interval to estimate a population mean, given a sample mean of $\overline{x} = 100$, a population standard deviation of $\sigma = 20$, and a sample size of $n = 1{,}000$.
4. Approximate the level of confidence assigned to the interval estimate $49 \leq \mu \leq 51$ for a population mean, given a sample mean of $\overline{x} = 50$, a population standard deviation of $\sigma = 8$, and a sample size of $n = 100$.

(Answers are on page 220.)

Population Proportions and Standard Error

The examples in this chapter so far have relied on knowing population standard deviations to construct confidence intervals about unknown population means. In practice, this isn't always possible. Sometimes you may know something about the population standard deviation. Maybe additional research or studies give you an idea of what it might be, but often the population standard deviation is just as unknowable as the population mean itself.

In cases where you don't know the population standard deviation, you might be able to estimate it from sample data. Such an estimate of standard deviation of the sample statistic is called the *standard error*. Determining a reasonable standard error for your sample isn't always possible or simple, but one situation where you can see how it works is in the case of population proportions and sample proportions, as explained in the next two sections.

Confidence Intervals for Population Proportions

Here's an example of constructing a confidence interval for a population proportion using the standard error in place of the unknown population standard deviation.

Example 4:

Imagine a mayoral race between Candidate A and Candidate B in a small city of 10,000 voters. A poll is conducted one month before the election, and 250 voters are asked if they will vote for Candidate A. The result of the poll is that 135 voters say yes and 115 voters say no. In Chapter 8, you learned that the sample proportion is an unbiased estimator of the population proportion. Since the sample proportion is $\frac{135}{250} = 0.54$, you estimate the population proportion to also be 0.54. In other words, you estimate that 54% of the population will vote for Candidate A.

This makes it seem like Candidate A has a good chance of being elected, but remember, sampling involves uncertainty. How much uncertainty is associated with this estimate?

As before, you can construct a confidence interval to quantify the uncertainty, again relying on the fact that under the right circumstances, the sampling distribution of the sample proportion can be approximated by the normal distribution. As you learned in the last section, to construct a 95% confidence interval, you just need to add and subtract two standard deviations from your estimate of 0.54. However, in the last section, you knew the population standard deviation. What do you do if you don't know that information?

Here's the formula for the standard deviation of the sampling distribution of sample proportions (where p is the population proportion and n is the sample size):

$$\sqrt{\frac{p(1-p)}{n}}$$

REMINDER

This formula should remind you of the formula for the standard deviation of the binomial distribution. As you learned in the last chapter, the distribution of sample proportions is a binomial distribution.

Unfortunately, this formula only really helps if you know the population proportion p, which you don't. In fact, that's precisely what you are trying to estimate in this example. This is where standard error comes in.

Standard Error and Sample Proportions

In the case of sample proportions, you can approximate the standard deviation of the sampling distribution by replacing the population proportion in the formula with the sample proportion, as follows:

$$\sqrt{\frac{\hat{p}(1-\hat{p})}{n}}$$

This is an example of standard error. With sample proportions, this formula for standard error is a good estimate of the sample standard deviation, as long as the sample size is large enough and the proportion isn't too close to 0 or 1. Those conditions are met in this example, so you can use the standard error to estimate the standard deviation of the sampling distribution:

$$\sqrt{\frac{\hat{p}(1-\hat{p})}{n}} = \sqrt{\frac{0.54(1-0.54)}{250}} = 0.032$$

With the standard deviation estimated, you can go ahead and compute your 95% confidence interval. The lower bound will be two standard deviations below the sample proportion, while the upper bound will be two standard deviations above the sample proportion:

$$0.54 - 2 \times 0.032 \leq p \leq 0.54 + 2 \times 0.032$$
$$0.476 \leq p \leq 0.604$$

This says that in 95% of samples of 250 voters, the proportion of voters choosing Candidate A will be between 0.476 and 0.604, so you are 95% confident that the true percentage of people who will cast their vote for Candidate A is between 47.6% and 60.4%. This is a less precise estimate than the original 54%, but it captures some of the uncertainty of the process. Candidate A's chances of being elected still look good, but of course, there's a lot of variability in that estimate.

As with the confidence intervals constructed earlier in this chapter, you can tighten your estimate by increasing the sample size, because variability in the sampling distribution decreases as the sample size increases. For example, if you sampled 500 voters and found the sample proportion to be 0.54, the standard error would be:

$$\sqrt{\frac{\hat{p}(1-\hat{p})}{n}} = \sqrt{\frac{0.54(1-0.54)}{500}} = 0.022$$

This results in a narrower 95% confidence interval of

$$0.496 \leq p \leq 0.584$$

In the case of sample proportions, computing the standard error is as simple as replacing the unknown population proportion p with the known sample proportion $\hat{p}$ in the formula for standard deviation. For other population parameters, the situation may not be as simple.

For example, in the case of sample means, a standard error formula exists for estimating the standard deviation of the sampling distribution, but applying that formula requires more advanced statistical tools than those that have been discussed here. You'll learn such tools if you continue your statistical studies.

CAUTION—Major Mistake Territory!

Don't assume that techniques that work for one kind of statistical problem will automatically work for another. What works for sample proportions doesn't always work for sample means!

BRAIN TICKLERS Set #29

1. Given a sample proportion of $\hat{p} = 0.75$ from a sample of size 100, construct a 95% confidence interval for the population proportion.
2. Given a sample proportion of $\hat{p} = 0.75$, how large a sample would you need for your 95% confidence level estimate of the population proportion to have a margin of error of 0.05?
3. Suppose an unfair coin shows tails with probability p. If the coin is flipped 100 times and tails comes up 60 times, give a 95% confidence interval for p.

(Answers are on page 220.)

Brain Ticklers—The Answers

Set #27, page 208

1. $49.5 \le \mu \le 50.5$
2. $\frac{12}{\sqrt{25}} = 2.4$
3. $\overline{x} - \frac{\sigma}{\sqrt{n}} \le \mu \le \overline{x} + \frac{\sigma}{\sqrt{n}}$
4. Solve $\frac{\sigma}{\sqrt{n}} = 1$ to get $n = 100$.

Set #28, page 215

1. $96.08 \le \mu \le 103.92$
2. $94.85 \le \mu \le 105.15$
3. $98.76 \le \mu \le 101.24$
4. Around 80% confidence

Set #29, page 219

1. $0.66 \le p \le 0.84$
2. $n = 300$
3. Here, p is the population proportion of tails, and the sample proportion is $\hat{p} = \frac{60}{100} = \frac{3}{5}$. The estimate of the standard deviation is $\sqrt{\frac{\frac{3}{5} \times \frac{2}{5}}{100}} = 0.05$, so the 95% confidence interval is $0.50 \le p \le 0.70$.

Chapter 10

Statistical Significance

In this chapter, you'll learn how to use statistics to evaluate and decide among different hypotheses. Is your company's bump in sales a result of last month's new ad campaign, or is it the result of a new product that just launched? Is the aspirin you took this morning the reason your headache is gone, or would it have gone away by itself? Questions like these lie at the intersection of probability and statistics; knowledge of statistical sampling can help you answer them. The tools of inference developed in the previous two chapters are key in determining whether or not an event can be described as "statistically significant."

Significance Testing

In Chapter 5, you studied normal distributions and used their properties to answer simple questions about heights. Assuming that the heights of adult American men are normally distributed with a mean of 70 inches and a standard deviation of 3 inches, you determined that roughly 84% of adult American men are less than 6 feet, 1 inch tall and nearly 98% are less than 6 feet, 4 inches. Only about 2 out of every 100 adult American men are taller than 6 feet, 4 inches, so that height is unusual, if not rare. It is both your knowledge of statistics (things like standard deviations and z-scores) and your knowledge of the population distribution (the fact that heights are normally distributed) that allow you to draw this conclusion.

You can use this same strategy to draw more sophisticated inferences from data through significance testing. A *significance test* evaluates a given hypothesis in light of the data. It tells you how likely your hypothesis is to be true, given your observations and assumptions. For this reason, significance testing is also known as *hypothesis testing*, and significance tests are sometimes referred to as *tests of significance*.

Defining Significance

If your hypothesis is that the average adult American man is 5 feet, 10 inches tall, then meeting a man who is 6 feet, 4 inches tall wouldn't be that unusual since you know that 2 out of every 100 men should be that tall. On the other hand, meeting many men that tall would be unusual. If you selected 100 men at random and 20 of them were 6 feet, 4 inches or taller, that would be quite surprising. It might make you question whether your hypothesis (that the average male height is 5 feet, 10 inches) is really true.

A significance test tells you just how unlikely an observation from a population is, given some hypothesis about the population. It indicates how "significant" an observation is in light of what you believe to be true. Significance testing is closely related to confidence intervals; thus, the process relies heavily on knowledge of sampling distributions, estimators, and population parameters—everything you learned about in the past two chapters.

> "Significance" is another familiar word that has a very specific meaning in the context of statistics. When you read "significance," it's all right to think of it as "important" or "worthy of note." Remember that in the context of statistical inference, though, it has a technical meaning (how unlikely an observation is given some hypothesis) that is different than what it means in other, less formal situations.

Here's an example that will help put the definition of significance into context.

Example 1:

Suppose you are studying high school students' SAT scores, which are known to be normally distributed, with a population mean of $\mu = 1000$ and a population standard deviation of $\sigma = 200$. You have a hypothesis: you think that students who participate in sports get higher than average SAT scores. Because there are a large number of students who participate in sports, their SAT scores are also normally distributed, but your hypothesis is that the mean of their normally distributed scores is higher than the mean of the entire population, which is 1000.

To test this hypothesis, you take a simple random sample of 25 high school students who all participate in at least one school sport and record their SAT scores. The mean SAT score of this sample is 1050. Does this result support your hypothesis? The obvious answer seems to be yes. The student athletes have an average SAT score of 1050, which is greater than the population average of 1000.

Remember, however, that statistical sampling always involves random variation. Although the population mean of SAT scores is 1000, that doesn't mean that every random sample you take from that population will have a sample mean of 1000. Sometimes the sample mean will be higher than 1000, and other times it will be lower. The sample mean is an unbiased estimator of the population mean, which tells you that the average of all the possible sample means is 1000. Because of random variation, though, any one sample mean might be quite different.

So, just how different is your sample mean of 1050? Is it a sample mean you might expect to observe because of random variation? Or is it so unusual that it causes you to rethink your assumption that the average overall SAT score is 1000? This is the distinction you can make using a significance test.

Testing for Significance

To begin a significance test, you must first identify the null hypothesis. The *null hypothesis* is a statement about a population parameter, and it can usually be thought of as what is conventionally assumed to be true, or the usual state of affairs. In Example 1, the null hypothesis is that the population mean of SAT scores of high school students who participate in at least one school sport is the same as the population mean of SAT scores of high school students overall.

Once you have stated the null hypothesis, you then state the alternative hypothesis. The *alternative hypothesis* is a statement about the same population parameter; that is, it indicates how you think reality might differ from what is generally believed to be true. In this example, the alternative hypothesis is that the population mean of SAT scores of high school students who participate in at least one school sport is higher than the overall average of 1000.

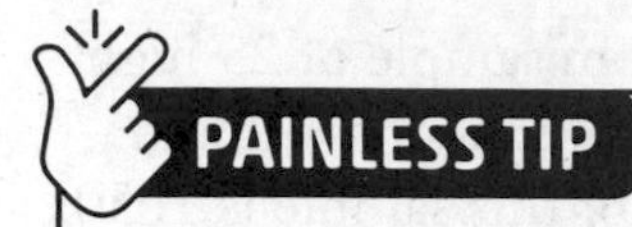

Significance tests often involve a comparison, and the null hypothesis is the claim that there is no difference between the things you are comparing. The alternative hypothesis is the claim that there is a difference between the things you are comparing and that difference is responsible for the unusual outcomes being observed.

The significance test asks this: given a statistical observation, how unlikely is that observation given that the null hypothesis is true? If the answer is "very unlikely," then your observation may be reason to reject the null hypothesis. If the answer is "fairly likely," then your observation is not a reason to reject the null hypothesis. For example, when considering the heights of adult American men, if 20% of the men you encounter during the course of a day are taller than 6 feet, 4 inches, then the null hypothesis—the assumption that men, on average, are 5 feet, 10 inches tall—might not be correct. This could give you reason to reject the null hypothesis. On the other hand, observing that 20% of the men you encounter in a day are 6 feet, 1 inch or taller would not be unusual under the assumption that the average man is 5 feet, 10 inches tall. This would not give you reason to reject this null hypothesis.

Drawing inferences about a population using statistical sampling requires that the samples be representative of the population. The easiest way to ensure this is to use simple random samples. You might meet a lot of unusually tall people at a professional basketball player's birthday party, but this wouldn't be a reason for you to change your opinion about the average height of a person because it's not a random sample of the population.

Turning back to the case of SAT scores from Example 1, the null hypothesis is that the population mean for student athletes is 1000;

that is, there is no difference between the two means (for student athletes and for high school students overall). Assuming this is true, then the mean of a sampling distribution of sample means should be 1000 as well because the sample mean is an unbiased estimator of the population mean. You now have a specific question about sampling distributions to investigate: In a sampling distribution of sample means with mean 1000 and sample size 25, how unlikely is a sample mean of 1050? The knowledge and techniques discussed in the previous two chapters can answer this question for you.

Since the population of SAT scores is assumed to be normally distributed, so too is the sampling distribution. You also know that the population standard deviation is $\sigma = 200$, so you can use this to compute the standard deviation of the sampling distribution for a sample of size $n = 25$:

$$\frac{\sigma}{\sqrt{n}} = \frac{200}{\sqrt{25}} = 40$$

Now you know everything you need to know about the sampling distribution of sample means. It is normal, has a mean of 1000, and has a standard deviation of 40. Whether an observed sample mean of 1050 is unusual or not depends on where that observation falls in this known distribution.

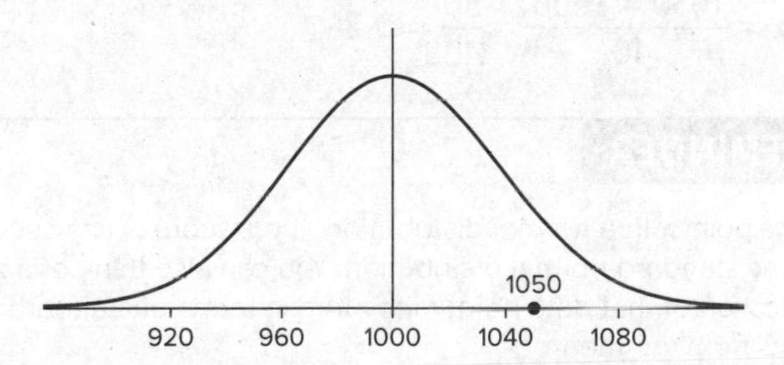

Figure 10–1. Sampling Distribution of Sample Means of SAT Scores

As you can see, a sample mean of 1050 is not that far from the mean in the sampling distribution. It is a little bit more than one standard deviation away. Given what you know about the spread of data in a normal distribution, you can see that this data point is not very unusual.

If observing a sample mean of 1050 was very unlikely in a population with a mean of 1000, this might give you reason to think that the assumption that the population mean is 1000 is incorrect. In other words, based on such an observation, you might reject the null hypothesis that the mean SAT score for student athletes is the same as that of students overall. However, the result of this significance test suggests that this observation is not so unusual, so it doesn't really constitute evidence against the null hypothesis. In this case, you have no reason to reject the claim that average SAT scores among students who participate in sports and students overall are the same.

This is the basic idea behind a significance test; statistical knowledge of normal distributions can help you draw conclusions that are more specific.

Levels of Significance and *p*-values

Fundamentally, a significance test tells you how surprised you should be by an observation. Of course, there are different levels of surprise, and this can be quantified using your knowledge of normal distributions.

In Example 1, an observation of 1050 in a distribution with a mean of 1000 and a standard deviation of 40 has a z-score of 1.25:

$$z = \frac{1050 - 1000}{40} = \frac{50}{40} = 1.25$$

REMINDER

For a data point *x* in a normal distribution, the z-score is its associated point in the standard normal distribution. You can also think of a z-score as the location of that data point measured in terms of standard deviations from the mean.

Using the standard normal table, you can see that about 89% of the data lies to the left of $z = 1.25$ in the standard normal distribution. This means that roughly 11% of the data lies to the right of this value.

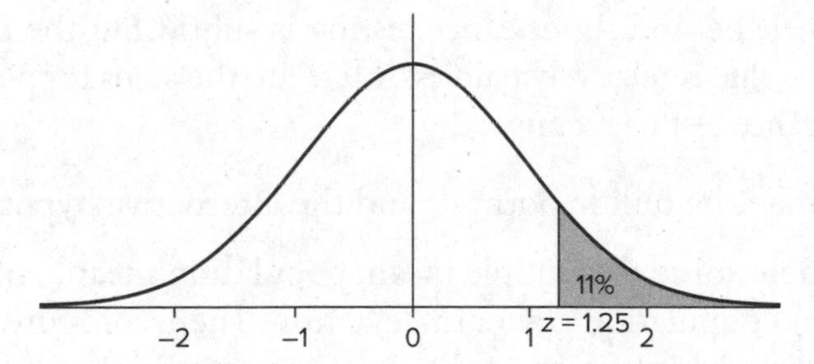

Figure 10–2. Standard Normal Distribution

This tells you that in the sampling distribution of sample means, about 11% of the sample means will be above 1050. If you were to repeatedly take random samples of size 25 from the population with a mean of 1000, about 11% of the time you would get a sample mean that is higher than 1050. This is around a 1 in 9 chance, which is not that unusual. This is why you probably wouldn't reject the null hypothesis based on this observation. The sample mean of 1050 was more likely the result of random variation than a higher than average population mean.

The 11% is an important number in the significance test. This probability is called the *p*-value of the observation. Small *p*-values are evidence against the null hypothesis because a small *p*-value means the observation was very unlikely given the null hypothesis. In this situation, a sample mean of 1050 has the *p*-value of 0.11, which is not that small, meaning the observation was not particularly unlikely.

PAINLESS TIP

The *p*-value of an observation is the probability a value as extreme as or more extreme than the observation will be seen in a sample from the population. The lower the *p*-value, the more surprising the observation is and the more likely you are to reject the null hypothesis.

Example 2:

Now suppose that, instead of 1050, your sample of SAT scores from student athletes had a mean of 1100. What is the *p*-value associated with this new observation? Should you reject the null hypothesis because of it?

The rationale behind significance testing is subtle, but the process of finding a p-value is relatively painless. Here are the steps for performing the significance test in Example 2.

Step 1: State the null hypothesis and the alternative hypothesis.

Step 2: Determine the sample mean, population mean, sample size, and population standard deviation. Then, compute $\frac{\sigma}{\sqrt{n}}$, the standard deviation of the sampling distribution.

Step 3: Compute the z-score by subtracting the population mean from your sample mean and dividing that result by the standard deviation.

Step 4: Locate the z-score (from Step 3) on the standard normal table. That value is the percentage of data that lies to the left of that z-score.

Step 5: Subtract that value (from Step 4) from 1. This value is the p-value of your observation. If the p-value is very small, reject the null hypothesis. If the p-value is not that small, do not reject the null hypothesis.

In this example, both the null hypothesis and the alternative hypothesis are the same as in Example 1. The null hypothesis is that the population mean of SAT scores of high school students who participate in at least one school sport is the same as the population mean of SAT scores of high school students overall. The alternative hypothesis is that the population mean of SAT scores of high school students who participate in at least one school sport is higher than the overall average of 1000.

Here in Example 2, the sample mean is 1100, the population mean is 1000, the sample size is 25, and the population standard deviation is 200. The standard deviation is $\frac{\sigma}{\sqrt{n}} = \frac{200}{\sqrt{25}} = \frac{200}{5} = 40$. Now, you can compute the z-score as follows:

$$\frac{1100 - 1000}{40} = \frac{100}{40} = 2.5$$

According to the standard normal table, 99.38% of the data in a normal distribution lies to the left of a z-score of 2.5. By the complement rule, less than 1% of the data lies to the right of that z-score. Therefore, in this case, a sample mean of 1100 has a p-value of less than 0.01, or 1%.

The complement rule tells you that if the probability of an event is p, then the probability of everything else is $1 - p$.

The sample mean of 1050 from Example 1 has a p-value of 0.11, and 11% doesn't seem surprising enough to reject the null hypothesis. However, a p-value of 0.01, or 1%, as seen here in Example 2, is much more surprising. This is a 1 in 100 chance, which is like flipping heads six times in a row on a fair coin. Observing this sample mean of 1100 might lead you to reject the null hypothesis that the scores of student athletes are the same as those of all students overall.

Testing for Specific Levels of Significance

It's common to test a hypothesis at a predetermined level of significance. In this situation, you decide on a level of significance beforehand and then compute the p-value of your observation. If the p-value of your observation is less than the level of significance, then the observation is called *statistically significant*. For example, the p-value of the sample mean of 1050 was 0.11. This is statistically significant at the 15% level (since 0.11 is less than 0.15) but not statistically significant at the 5% level (since 0.11 is greater than 0.05).

Reasonable, and important, questions to ask in the context of significance testing are: Where do you draw the line between surprising and unsurprising? At what point do the results of the significance test transition from "not surprising enough to reject the null hypothesis" to "surprising enough to reject the null hypothesis"? There are no definite answers. In many situations where statistics is applied, the convention is that the cutoff is 5%. A result with a p-value less than 0.05 is called "statistically significant" and is taken as evidence to reject the null hypothesis, while a result with a p-value greater than 0.05 is not considered "statistically significant" nor is it considered evidence against the null hypothesis.

However, this cutoff of 5% is an arbitrary choice. Is a p-value of 0.04 really that different than a p-value of 0.06? Both observations are similarly unlikely under the assumption of the null hypothesis, but one is considered "significant" and the other is not. You'll encounter the notion of significance in your studies and applications of statistics. It's important to remember that in the subtle and complex process of hypothesis testing, the level at which something becomes "significant" is a matter of convention. Using the word is not a substitute for understanding the process.

It's also worth acknowledging that a 5% significance level, which is frequently the cutoff for scientific studies, doesn't actually mean your observation is especially unusual. A p-value of 0.05 means that an observation as extreme, or more extreme, occurs roughly 5% of the time. That's a 1 in 20 chance, comparable to flipping heads four times in a row. Or in another example, if you performed an experiment 100 times, you'd expect such an event to occur 5 times. This may be unlikely, but it's certainly not unusual.

> It is common to describe scientific findings as "statistically significant," but it is important to remember that what counts as "significant" is a convention. A significance level of 5% is often chosen to be the standard, but some statisticians and scientists argue for a lower cutoff for significance. Others even argue against the notion of significance altogether. And remember, even at a significance level of 5%, a "statistically significant" observation will still occur 1 of out every 20 times just by random chance.

BRAIN TICKLERS Set #30

1. SAT scores are normally distributed with a population mean of $\mu = 1000$ and a population standard deviation of $\sigma = 200$. What percentage of SAT scores should be

   a. higher than 1200?

   b. lower than 750?

   c. more than 200 points different than the mean?

   d. higher than 1000?

2. A researcher wants to test the impact that eating sugary breakfast cereal has on one's blood pressure. Give an example of a possible null hypothesis and an alternative hypothesis for a significance test for this experiment.

3. A researcher who is studying corporate salaries has a hypothesis that employees who graduated from elite colleges have higher salaries, on average, than employees overall. If the average corporate salary is $50,000 with a standard deviation of $20,000, would a sample mean of $55,000 from 100 employees who graduated from elite colleges be statistically significant at the 5% level?
4. If you flipped a coin 10 times and saw 7 heads, would you think the coin was unfair? Would your opinion change if you flipped that coin 100 times and saw 70 heads?

(Answers are on page 245.)

One-Sided and Two-Sided Tests

As a result of the important role the normal distribution has in sampling, significance testing often involves computing z-scores. In Examples 1 and 2, the z-score was the *test statistic*, the statistic you used to test your hypothesis about the population parameter. The *z-test statistic* takes the following form (where $\overline{x}$ is the sample mean, μ is the population mean, and $\frac{\sigma}{\sqrt{n}}$ is the standard deviation of the sampling distribution):

$$z = \frac{\overline{x} - \mu}{\frac{\sigma}{\sqrt{n}}}$$

As you'll see in this section, the way the test statistic is used depends on whether you are performing a one-sided or two-sided test.

One-Sided Significance Tests

In Example 1, the significance test required computing the probability of observing a sample mean greater than or equal to 1050 in a random sample taken from a population of mean 1000, so you computed the z-test statistic using $\overline{x} = 1050, \mu = 1000$, and $\frac{\sigma}{\sqrt{n}} = \frac{200}{\sqrt{25}} = 40$:

$$\frac{\overline{x} - \mu}{\frac{\sigma}{\sqrt{n}}} = \frac{1050 - 1000}{\frac{200}{\sqrt{25}}} = \frac{50}{40} = 1.25$$

The significance test came down to the question "What is the probability of seeing a z-score this high, or higher, in the standard normal distribution?" The answer to this question is given by:

$$P(Z \geq 1.25)$$

This is the probability that Z, a randomly selected value from the standard normal distribution, is greater than or equal to 1.25. From the standard normal table, you know that $P(Z \leq 1.25) = 0.8944$, and by the complement rule, $P(Z \geq 1.25) = 1 - P(Z \leq 1.25) = 1 - 0.8944 = 0.1056$. This corresponds to a p-value of 0.106, and so the probability that a randomly selected value from the standard normal distribution will be greater than or equal to 1.25 is around 10.56%, or roughly 11%, of the time.

This is an example of a *one-sided* (or *one-tailed*) hypothesis test. Observations that are statistically significant occur on only one side, or in one tail, of the normal distribution. You can tell that a one-sided test is called for in this situation because of your hypothesis. The null hypothesis—the "no difference" hypothesis—was that the average SAT scores of student athletes is no different than the average SAT score of students in general. That is:

$$H_0 : \mu_a = \mu$$

Here, the symbol μ_a, read "mew sub *a*," represents the population mean of SAT scores of student athletes. So the equation $\mu_a = \mu$ says that the population mean of SAT scores of student athletes is equal to the population mean of overall SAT scores.

The alternative hypothesis in this case is that the average SAT scores of student athletes is higher than that of students in general:

$$H_a : \mu_a > \mu$$

1+2=3 MATH TALK!

The symbol for the null hypothesis, H_0, is read "*H* nought." (In other mathematical contexts, you might read as "*H* sub 0," but in statistics, you usually read it as "*H* nought.") The symbol for the alternative hypothesis, H_a, is read "*H* sub *a*," where the "sub" refers to "subscript."

Because of the way the alternative hypothesis is written, only evidence that suggests the population mean is greater than 1000 will be considered significant. So when your sample mean comes in at 1050, you are only interested in the probability that a sample mean could be that extreme or higher. Visually, this means you are interested in a region such as this.

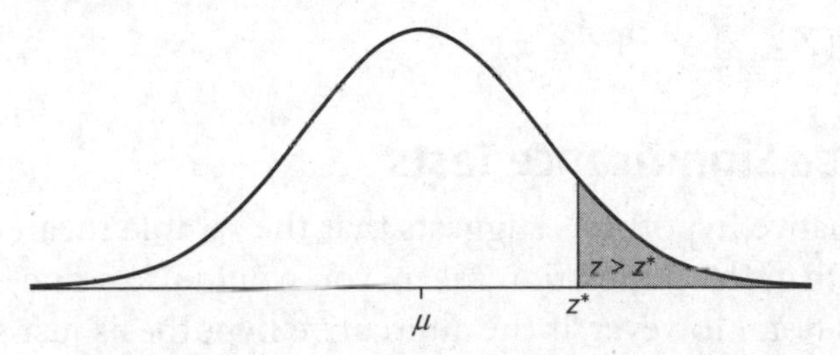

Figure 10–3. A One-Sided (High Side) Significance Test

The shaded region consists of data that is far from the mean and above it. It is these observations that would suggest to you that your alternative hypothesis—a higher SAT score population mean for student athletes—might be true. For example, an observed sample mean of 1100 has a z-score of 2.5, which puts it far to the right of the normal distribution. The shaded region in the following figure corresponds to observations that are at least 2.5 standard deviations above the mean.

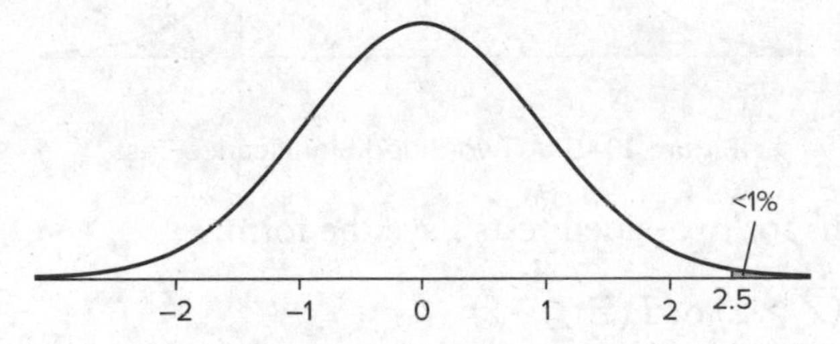

Figure 10–4. Standard Normal Distribution

The shaded region in this figure comprises less than 1% of the overall area of the normal distribution, and so an observation this extreme occurs less than 1% of the time. This results in a p-value of less than 0.01.

The test here is one sided, or one tailed, in the sense that only a region on one side of the normal distribution would provide evidence to reject the null hypothesis. You are looking for evidence that the observed mean is both far from and above the population mean. Since you are only interested in the probability that you see an observation as high as, or higher than, the upper critical value z, this is referred to as a one-sided *high side test*. You can also perform a one-sided *low side test*. These probabilities take the form:

$$P(Z \geq z) \text{ or } P(Z \leq z)$$

Two-Sided Significance Tests

If an alternative hypothesis suggests that the sample mean is higher (or lower) than the population mean, you would use a one-sided hypothesis test. However, if the alternative hypothesis just states that the two means are different (without specifying which is higher or lower), then you would use a *two-sided* (or *two-tailed*) hypothesis test. In a two-sided test, the process is the same as for a one-sided test, but now an extreme observation on either the high side or the low side would constitute evidence against the null hypothesis.

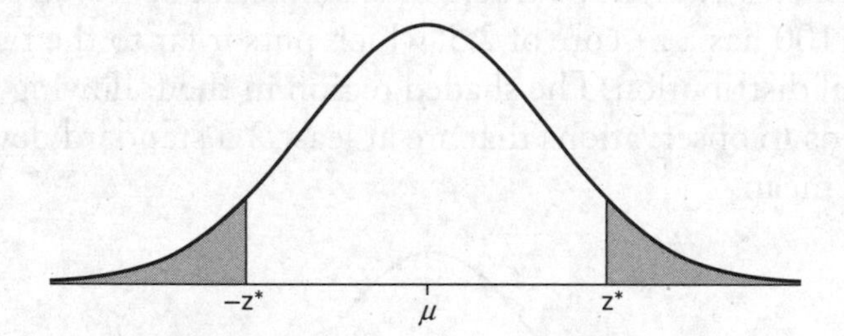

Figure 10–5. A Two-Sided Significance Test

Calculations for two-sided tests have the form:

$$P(Z \geq z) \text{ or } P(Z \leq -z)$$

This is because in a two-sided test, you need to consider extreme observations in both directions (both above and below the mean). In a two-sided test, an unusually low observation would be just as significant as an unusually high observation. Both would count as evidence against the null hypothesis.

The symmetry of the normal distribution makes calculations for two-sided tests easy. Since two opposite z-scores, z and $-z$, are symmetric about the mean in a standard normal distribution, the regions they determine are equal in area. You can see this symmetry in the standard normal table itself. Consider the z-score of 2. The area under the standard normal curve to the right of $z = 2$ is:

$$P(Z \geq 2) = 1 - P(Z \leq 2)$$

You can use the standard normal table to compute this. The entry for $P(Z \leq 2)$ is 0.9772, so $P(Z \geq 2) = 1 - 0.9772 = 0.0228$. Now, the area to the left of $z = -2$ is given by $P(Z \leq -2)$, which can be read directly from the standard normal table and is 0.0228. You can see that $P(Z \geq 2) = P(Z \leq -2)$.

In general, $P(Z \geq z) = P(Z \leq -z)$ for any z-score, z, due to the symmetry of the standard normal distribution. So instead of computing both $P(Z \geq z)$ and $P(Z \leq -z)$ and adding them, you can compute one of them and double it. You can express this relationship with the following equation:

$$P(Z \geq z) + P(Z \leq -z) = 2P(Z \geq |z|)$$

REMINDER

Taking the absolute value of a number will never produce a negative result. In the case of a z-score, $P(Z \geq |z|)$ guarantees that you are computing the area of the upper region of the appropriate two-sided tail.

Here's an example of a two-sided significance test at a fixed confidence level.

Example 3:

Imagine a company that purchases bolts from a supplier for use in one of their commercial products. The supplier's manufacturing process is expected to produce bolts that are 5 centimeters in diameter, and the process is known to have a standard deviation of 0.05 centimeter, which is an acceptable margin of error for the company's purposes. Every month when an order of bolts arrives, the company selects 10 bolts at random and measures their diameters to make sure the products are up to specifications. If the average diameter is 5 centimeters at the 5% significance level, the bolts will be rejected and sent back to the supplier.

The null hypothesis is that $\mu = 5$, and the alternative hypothesis is that $\mu \neq 5$. In this case, the company purchasing the bolts would perform a two-sided test. If the average bolt diameter is either too high or too low, the bolts will be rejected. This month's sample mean is 5.04 centimeters, but this observation has to be checked to see if it is too high or too low.

For a two-sided test at a 5% significance level, half of the extreme observations will occur on the high side and half will occur on the low side. This means you need to find the upper critical value for 2.5%, or 0.025. According to the standard normal table, this corresponds to a z-score of $z = 1.96$. If the absolute value of the z-test statistic is greater than $z = 1.96$, then the observation will be statistically significant at the 5% level.

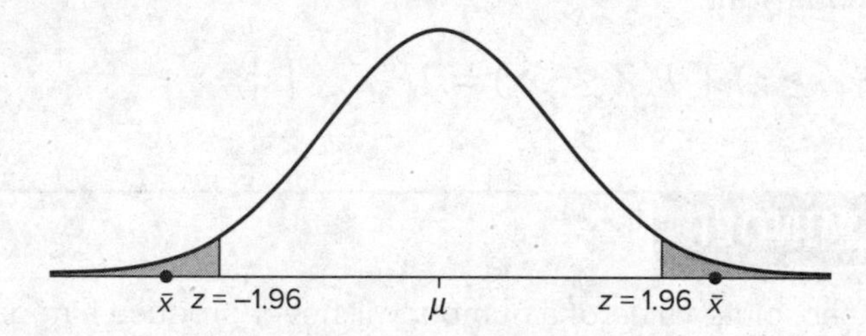

Figure 10–6. Statistically Significant Observations at the 5% Level

Assuming the null hypothesis is true, $\mu = 5$, and so you can compute the z-test statistic using the assumed population mean and known sample mean, standard deviation ($\sigma = 0.05$), and sample size ($n = 10$):

$$z = \frac{\bar{x} - \mu}{\frac{\sigma}{\sqrt{n}}} = \frac{5.04 - 5}{\frac{0.05}{\sqrt{10}}} = 2.53$$

The z-test statistic of 2.53 is larger than the cutoff of 1.96, so this observation is statistically significant at the 5% level. Thus, there is reason to reject the null hypothesis that the mean diameter of all the bolts is 5 centimeters because there is less than a 5% chance that these bolts were so unusual due to chance alone. This company would send these bolts back to their supplier.

The key to performing a one-sided or two-sided significance test at a given level of confidence, α, is determining the appropriate critical values for your test. For a one-sided high side test, where the statistically significant observations appear above the cutoff, the critical value is the z-score z^* such that $P(Z \geq z^*) \leq \alpha$. For a *one-sided low side* test, where the statistically significant observations appear below the cutoff, the critical value is the z-score z^* such that $P(Z \leq z^*) \leq \alpha$. For a two-sided test, half of the statistically significant observations occur on the high side and half occur on the low side. Thus, the critical value you want is the z^* so that $P(Z \geq |z|) \leq \frac{\alpha}{2}$.

In a two-sided significance test at a level of significance α, observations that suggest rejecting the null hypothesis are exactly those observations that fall outside a $1 - \alpha$ confidence interval for the mean. For example, in the standard normal distribution, a 95% confidence interval for the mean is $-1.96 \leq z \leq 1.96$, as this interval covers 95% of possible values. In Example 3, the observation of $\bar{x} = 5.04$ cm corresponds to a z-score of 2.53, which lies outside this interval and thus in the other 5% of values. This makes the observation significant at the 5% level.

BRAIN TICKLERS Set #31

1. For each of the following situations, indicate whether you would use a one-sided or two-sided significance test.
   a. Analyzing whether left-handed shooters in basketball have a higher shooting percentage
   b. Analyzing whether people with mathematics degrees can more accurately estimate grocery bills
   c. Analyzing whether adults who run three times per week have lower cholesterol
2. Given an observed sample mean of $\bar{x} = 115$ from a sample of size 10 drawn from a population with mean μ and $\sigma = 20$, will the result of a one-sided (high side) significance test at the 5% significance level suggest that you reject the null hypothesis that $\mu = 100$?
3. Given an observed sample mean of $\bar{x} = 48$ from a sample of size 25 drawn from a population with mean μ and $\sigma = 3$, will the result of a two-sided significance test at the 5% significance level suggest that you reject the null hypothesis that $\mu = 50$?

4. Explain how the same observation might be statistically significant at the 5% level in a one-sided test but not statistically significant at the 5% level in a two-sided test.

(Answers are on pages 245–246.)

Type I and Type II Errors

The language of hypothesis testing is subtle and technical. An observation that is very unlikely due to random variation alone is considered evidence to "reject the null hypothesis." Notice that a significance test doesn't determine whether or not a hypothesis is true; it just evaluates a piece of evidence for or against a hypothesis based on how surprising it is in the context of random sampling.

Even though a significance test doesn't tell you whether a hypothesis is true or false, significance testing is still used in decision making. The results of a test might lead to you to decide to make a policy decision or send back a batch of defective products. However, as a result of the uncertainty inherent in the process of significance testing, erroneous conclusions will sometimes be drawn, and there are two types of errors that can be made.

The first type of error is when you reject the null hypothesis even though it is true. This is known as a *type I* error. The second type of error is when you fail to reject the null hypothesis even though it is false. This is known as a *type II* error. The examples that follow demonstrate these types of errors.

Computing the Probability of a Type I Error

Example 4:

Imagine you decide to test your hypothesis that student athletes have higher than average SAT scores by choosing a significance level of 5%, or $\alpha = 0.05$, and performing a one-sided (high side) significance test. According to the standard normal table, $P(Z \geq 1.65) = 0.05$, so the upper critical value for your test is $z = 1.65$, and the z-test statistic for a sample mean of 1050 is:

$$\frac{\bar{x} - \mu}{\frac{\sigma}{\sqrt{n}}} = \frac{1050 - 1000}{\frac{200}{\sqrt{25}}} = \frac{50}{40} = 1.25$$

In a one-sided (high side) significance test, you reject the null hypothesis if the z-test statistic is greater than the upper critical value. Here, the z-test statistic of 1.25 is less than 1.65, which means that this observation is consistent with the null hypothesis being true, that is, that the mean SAT scores of both groups are the same. As a result of this, you do not reject the null hypothesis.

On the other hand, an observation corresponding to a z-test statistic higher than 1.65 would lead you to reject the null hypothesis. However, there's a chance that this would be a mistake. Think back to the theory of statistical sampling. Assuming the null hypothesis is true, the 5% significance level means that in 95% percent of the samples you could draw from the population, the z-test statistic associated with the sample mean will be less than 1.65. But that also means that in 5% of the samples you can take from the population, the z-test statistic will be above 1.65 simply because of random variation. Such an observation would lead you to reject the null hypothesis even though it is true. This is a type I error.

Because of random variation, sample means will occasionally be unusual enough to suggest the null hypothesis is false even though it is true. Even if the population mean of student athlete SAT scores really is 1000, random sample means will occasionally be so high that they suggest the population mean is greater than 1000. At a 5% significance level, this will happen around 5% of the time.

PAINLESS TIP

In significance testing, a type I error—the error of rejecting the null hypothesis even though it is true—has the same probability as the level of significance. A type I error will occur *P*% of the time when testing for significance at the *P*% level.

Computing the Probability of a Type II Error

It is slightly more complicated to compute the probability of a type II error. A type II error occurs when you accept the null hypothesis even though it is false. This can happen when you fail to reject the null hypothesis even though an alternative hypothesis, one that is incompatible with the null hypothesis, is in fact true. To analyze this situation, you need a specific alternative hypothesis to test it against, as seen in the next example.

Example 5:

Suppose you have a specific alternative hypothesis that the true average SAT score of student athletes is 1100. That is:

$$H_a : \mu_a = 1100$$

In the original significance test at the 5% significance level, a z-test statistic of 1.65 or above would have been cause to reject the null hypothesis, so you can think of a z-test statistic below 1.65 as a reason to accept the null hypothesis. You can use the z-score calculation in reverse to determine what SAT score corresponds to a z-score of 1.65:

$$z = \frac{\bar{x} - \mu}{\frac{\sigma}{\sqrt{n}}}$$

$$1.65 = \frac{\bar{x} - 1000}{\frac{200}{\sqrt{25}}}$$

$$\bar{x} = 1.65 \times 40 + 1000$$

$$\bar{x} = 1066$$

Thus, if you observe a sample mean below 1066, you would accept the null hypothesis that the population mean is 1000.

If the population mean is in fact 1100, then accepting the null hypothesis would be a type II error. In this situation, the probability of committing this kind of error is exactly the probability of observing a sample mean less than 1066 when the population mean is 1100. To determine this probability, you need to find the new z-score associated with 1066 under the assumption that the population mean is now $\mu = 1100$:

$$z = \frac{\bar{x} - \mu}{\frac{\sigma}{\sqrt{n}}} = \frac{1066 - 1100}{\frac{200}{\sqrt{25}}} = -0.85$$

Thus, when the population mean is actually 1100, the probability of observing a sample mean less than 1066 is equal to the probability of observing a value less than −0.85 in the standard normal distribution. This is given by $P(Z \leq -0.85)$ and the value can be read off the standard normal table: $P(Z \leq -0.85) = 0.1977$. Thus, there is roughly a 20% chance that, even when the alternative hypothesis is true, you would accept the null hypothesis instead, resulting in a type II error. This situation is often illustrated with the following picture.

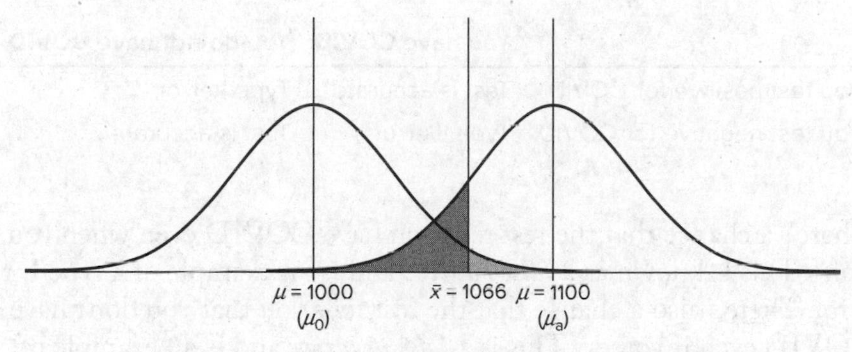

Figure 10–7. Illustration of Type I and Type II Errors

On the left is the normal distribution associated with the null hypothesis, and on the right is the normal distribution associated with the alternative hypothesis. The light gray region represents observations that would lead you to reject a hypothesis that is true, so the area of this region is the probability of a type I error. The dark gray region represents observations that would lead you to accept the null hypothesis even though the alternative hypothesis is true, so the area of this region is the probability of a type II error.

> The *power* of a statistical test against an alternative is the probability that the test will reject the null hypothesis when the alternative hypothesis is true. In other words, the power of a test against an alternative is the probability that the test will not result in a type II error. This makes the power of a test equal to 1 – the probability of a type II error.

These two kinds of errors are very different in nature, but they are both a result of random variation in sampling. Type I and type II errors also come up in other situations, and a simple way to understand them is in the context of medical testing.

Example 6:

Suppose you are experiencing flu-like symptoms and decide to get a COVID test. You might have COVID or you might not, and the test could come back positive or negative. Therefore, there are four possible situations, as outlined in the following table.

	You have COVID	You do not have COVID
You test positive for COVID	Test is accurate	Type I error
You test negative for COVID	Type II error	Test is accurate

There's a chance that the test tells you have COVID even when you don't. This is known as a *false positive* and is an example of a type I error. There's also a chance that the test tells you that you don't have COVID even if you do. This is a *false negative* and is an example of a type II error. You may recognize a similarity between this table and the two-way tables used in conditional probability in Chapter 7. It is no coincidence, as there is much conditional probability underlying the theory of significance testing and type I and type II errors. Further study of probability and statistics will lead you to these deeper connections.

CAUTION—Major Mistake Territory!

Don't make the mistake of assuming that type I and type II errors are equally likely. They are different kinds of errors with different probabilities and very different consequences. In the context of medical testing, a false negative (a type II error) is potentially more dangerous than a false positive (a type I error). This is because a false positive would likely trigger more testing that could resolve the uncertainty about a patient's status, while a false negative might make it seem like a patient is healthy when that's not the case. Medical tests often have a higher probability of false positives than false negatives because, statistically (and medically) speaking, it's better to be safe (to test positive when you don't have the disease) than sorry (to test negative when you do).

Be Cautious When Using Statistical Significance

There are many different kinds of errors that can result in significance testing. You just learned about type I and type II errors, which are consequences of the process of random sampling. These errors can be understood mathematically, and the uncertainty around them can be quantified.

Perhaps more dangerous are the errors associated with improperly applying and interpreting tests of significance. The significance tests presented in this chapter cannot be used on all types of data. Remember, in general, any process that relies on statistical sampling requires certain assumptions about the data—a finite mean, a finite standard deviation, simple random samples, a large enough sample size, a population much larger than the sample—to guarantee enough knowledge about the sampling distribution to be useful. In addition, the tests outlined here only apply when the population itself is normally distributed. If the population is not normal, different statistical tools and a more advanced knowledge of statistics will be required.

Perhaps the most important lesson you should take away from this chapter is understanding what statistical significance is and what it is not. In summary, statistical "significance" is a specific claim about a sampling distribution. It tells you how likely it would be for that unusual observation you see to appear due to chance alone. If such an observation is very unlikely to be produced by random variation, then it is statistically significant. That doesn't automatically mean it's important. You'll learn about some of the dangers of misinterpreting statistical significance in Chapter 12.

BRAIN TICKLERS Set #32

1. For each example, indicate whether it's a type I or type II error.
   a. A capable driver is denied a license because of a failing score on a written driving test.
   b. During a routine screening, a healthy individual tests positive for colon cancer.
   c. A message from your friend is marked as spam by your email system.
2. A hypothesis test is performed to determine if a dietary supplement helps people lose weight. What would be a type I error in this situation? What would be a type II error?
3. An observation is shown to be statistically significant at the 1% level. What is the probability of a type I error?

(Answers are on page 246.)

Brain Ticklers—The Answers

Set #30, pages 230–231

1. a. Around 15.8%

   b. Around 11.6%

   c. Around 31.7%

   d. 50%

2. A possible null hypothesis could be, "Eating sugary breakfast cereal has no impact on blood pressure," and an alternative hypothesis could be, "Eating sugary cereal raises blood pressure."

3. Yes. The z-score associated with the sample mean of $55,000 is $\frac{55{,}000 - 50{,}000}{\frac{20{,}000}{\sqrt{100}}} = \frac{5{,}000}{2{,}000} = 2.5$, and the probability of seeing a value 2.5 standard deviations or higher above the mean in a normal distribution is less than 1%.

4. In 10 coin flips of a fair coin, 7 heads isn't that unusual. However, in 100 flips of a fair coin, 70 heads would be very unusual. The larger sample size gives more weight to the hypothesis that the coin is unfair.

Set #31, pages 237–238

1. a. One-sided (high side)

   b. Two-sided

   c. One-sided (low side)

2. In this case, $\frac{115 - 100}{\frac{20}{\sqrt{10}}} = 2.37$, and since $P(Z \leq 2.37) = 0.991$, this observation has a p-value of less than 1%, making it statistically significant at the 5% level. This would suggest rejecting the null hypothesis.

3. The critical value here is $z^* = 1.96$ since $P(Z \geq 1.96) = P(Z \leq -1.96) = 0.025$. Since $z = \frac{48 - 50}{\frac{3}{\sqrt{25}}} = -3.33$ and $-3.33 < -1.96$, this result is statistically significant, and you would reject the null hypothesis.

4. In a two-sided test at the 5% level, 2.5% of the statistically significant values lie on each side of the mean. In a one-sided test, the entire 5% lies on one side of the mean, so the region will be bigger and include more values on that side.

Set #32, page 244

1. a. Type II error (False negative)

   b. Type I error (False positive)

   c. Type I error (False positive)

2. A type I error would be concluding that the supplement has an effect when it actually doesn't. A type II error would be concluding that the supplement has no effect when it actually does.

3. 1%

Chapter 11

Bivariate Statistics

The majority of this book has focused on single-variable data. So far, you have learned techniques for measuring the center, spread, and shape of single-variable data, how to apply knowledge of distributions, and how to draw inferences about populations using statistical sampling.

When it comes to analyzing two-variable, or *bivariate*, data, many of the same basic concepts apply. Visualizations can help you understand the shape of bivariate data, and inferences can be drawn under the right circumstances. Perhaps the biggest difference is that, with bivariate data, you are looking for relationships between the two individual variables. There is a wide and deep study of two-variable data. Some of the statistical techniques with two-variable data require advanced mathematics, but the basics of bivariate data covered in this chapter are painless.

The Basics of Bivariate Data

Single-variable, or univariate, data is a single piece of information about individuals in a population. Two-variable, or bivariate, data is a pair of data values from each member of the population. Bivariate data can be represented as ordered pairs of values, where each coordinate is a value from each of the individual variables.

An ordered pair is two values represented as a point (x, y).

You can apply all the tools you've learned for analyzing single-variable data to each individual variable independently, but with bivariate data you can do even more. You can search for trends and patterns in the data and hope to use knowledge of one of the variables to better understand the other.

When you begin studying a two-variable data set, you should consider the same questions as when you start with a single variable: What population is represented by the data? And, how are the variables defined? But you'll ask more questions when studying bivariate data. In particular, you'll question how the variables are related to one another. Sometimes, the answer might be that they aren't related. When there is a relationship, however, you can apply a wealth of statistical tools to analyze and understand the connections between the data.

Scatterplots

Just as with single-variable data, the easiest way to start making sense of bivariate data is to graph it. The most common visualization of bivariate data is a scatterplot. As you learned in Chapter 2, a scatterplot is a graph of bivariate data where each ordered pair is plotted in two dimensions. One of the variables is represented along the horizontal axis (the x-axis), and the other variable is represented along the vertical axis (the y-axis). Like all good data visualizations, scatterplots can help you identify patterns in data.

In Chapter 2, you saw this scatterplot of city population versus police budget.

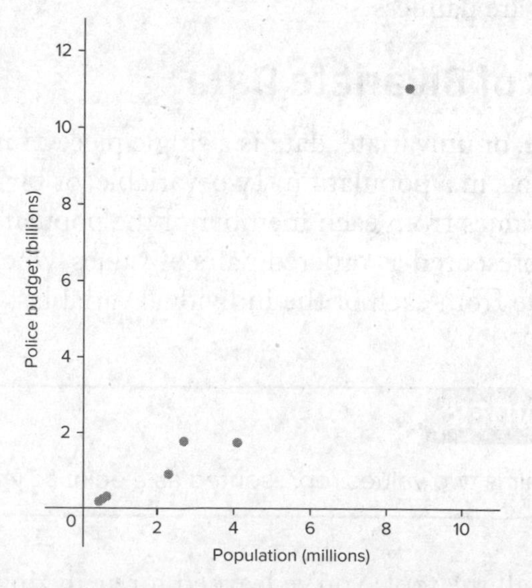

Figure 11–1. Police Bugdet vs. Population

This scatterplot makes it easy to spot a simple pattern in the data: bigger cities have larger police budgets. This probably makes sense, but it's also somewhat vague. Trying to pin down this pattern in a more precise way is one of the goals of analyzing bivariate statistics. You might use the data to try to understand a more specific question, such as "If City A has 500,000 more people than City B, how much larger should City A's police budget be?"

The scatterplot above makes other patterns visible. There are clusters of points in the lower left-hand corner. These are cities that are close in both population and police budget, and the clustering of the data might inspire you to look for other similarities among these cities. Also, there is a lone point far away from the data; as stated in Chapter 2, that point is New York City. It is both considerably larger than the other cities and its police budget is considerably higher. Despite being so far away from the other data, does New York City fit the overall pattern, or is it an outlier?

An outlier is a data point that does not fit the trend or the pattern of data. This is the same concept in bivariate data as it is in single-variable data.

Many scientific studies involve bivariate data. Frequently in these situations, one variable is considered the *response variable* and the other is the *explanatory variable*. The ultimate goal is to try to understand the influence of the explanatory variable on the response variable. Here's an example that shows how you can investigate the relationship between these kinds of variables.

Example 1:

Suppose that a teacher who posts optional review videos for students to watch is interested in whether or not those videos impact student performance. The teacher could create a bivariate data set where one variable is how many hours of videos each student watches and the other variable is the student's exam score.

The teacher could then plot this data in a scatterplot. In this scenario, "Hours of Videos Watched" is the explanatory variable, and "Exam Score" is the response variable.

PAINLESS TIP

As seen in this figure, the explanatory variable is usually plotted on the horizontal axis, and the response variable is plotted on the vertical axis.

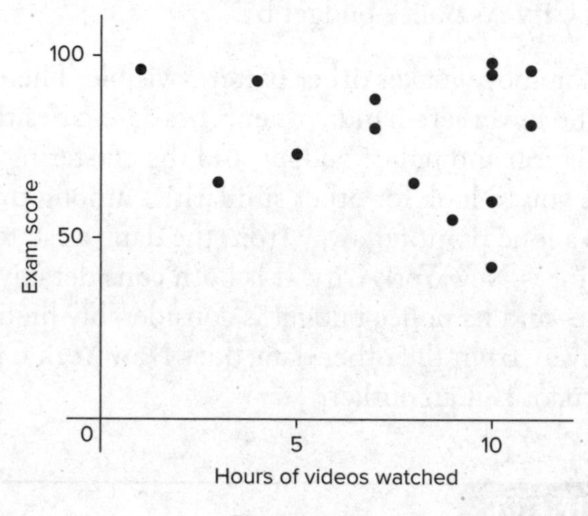

Figure 11–2

The goal of such an analysis is to understand the impact of the explanatory variable on the response variable. As the explanatory variable increases in value, what happens to the response variable? It might go up or down, and if it does, then the next task is to try to quantify that increase or decrease. Of course, it might be the case that there is no discernible pattern in the data.

In this example, a natural question to ask is "Do students who watch more review videos get higher exam scores?" As you can see, there are many points in the upper right part of the scatterplot. These are students who watched many hours of videos and received high scores on the exam. However, there are points in many different regions of the scatterplot. In the upper left-hand corner are students who watched very few hours of videos but scored high on the exam, and in the midright are students who watched many hours of videos but earned a midrange score on the exam.

Scatterplots can help you visualize bivariate data, but to make sense of what you are looking at, you need tools and vocabulary to help describe the relationship between the variables. When analyzing single-variable data, your first goal is to determine the center, spread, and shape of the data. These features help you understand the information you are working with. With bivariate data, your first goal is to determine the direction and association, form, and strength of the relationship between the variables.

Direction and Association of Bivariate Data

You read the *direction* of a scatterplot like you read a book. As you move from left to right along the horizontal axis, the data in the scatterplot might trend upward or downward, or it may have no apparent direction at all.

Here are some examples of scatterplots that have trends in an upward direction:

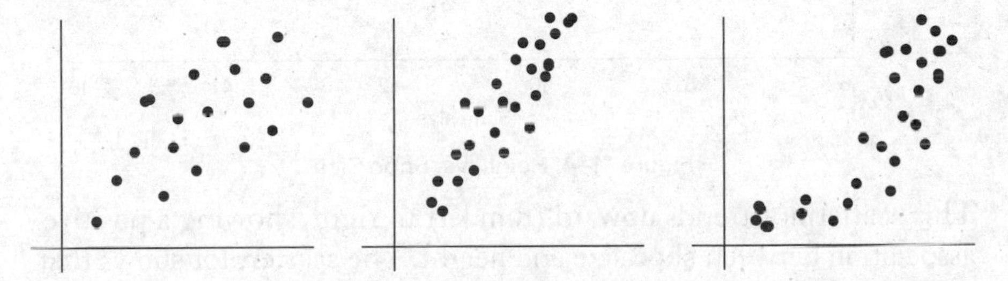

Figure 11–3. Positively Associated Data

These scatterplots show a *positive association* between the two variables. In bivariate data that is positively associated, high values of one variable are associated with high variables of the other. Likewise, low values of one variable are associated with low values of the other. Positively associated data trends upward as your read the graph from left to right.

A classic example of positively associated bivariate data is height versus shoe size, as illustrated in the following figure.

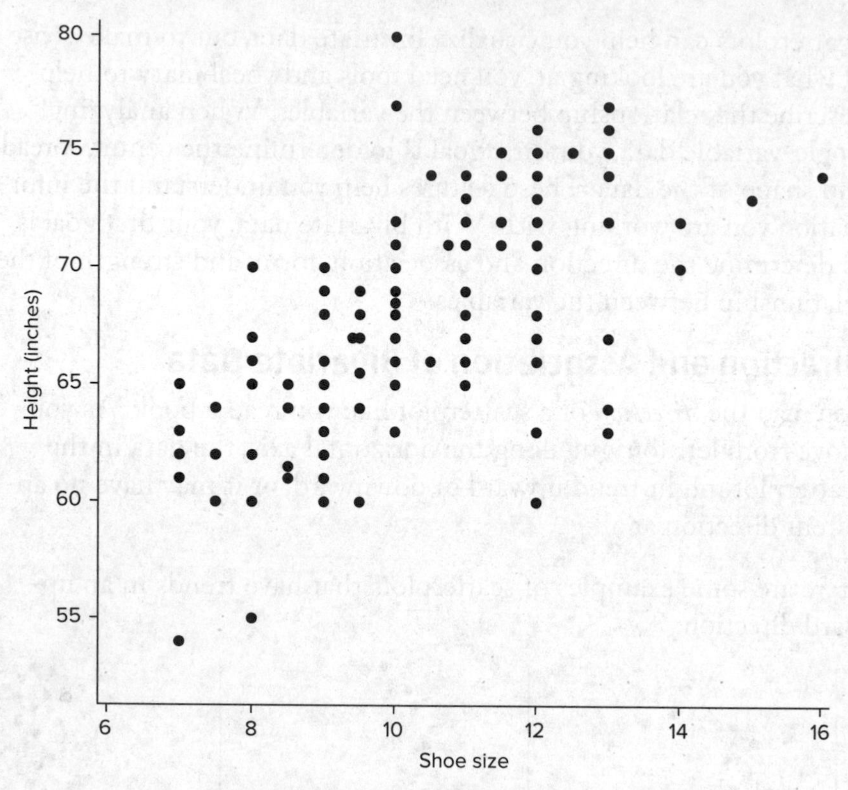

Figure 11–4. Height vs. Shoe Size

This scatterplot trends upward from left to right, showing a positive association between shoe size and height. The scatterplot shows that taller people are typically associated with bigger feet (and therefore a larger shoe size), and shorter people are typically associated with smaller feet. This relationship between the variables makes intuitive sense, as physical characteristics like height and shoe size are typically linked for each person.

Bivariate data that trends downward as you read the graph from left to right is *negatively associated.*

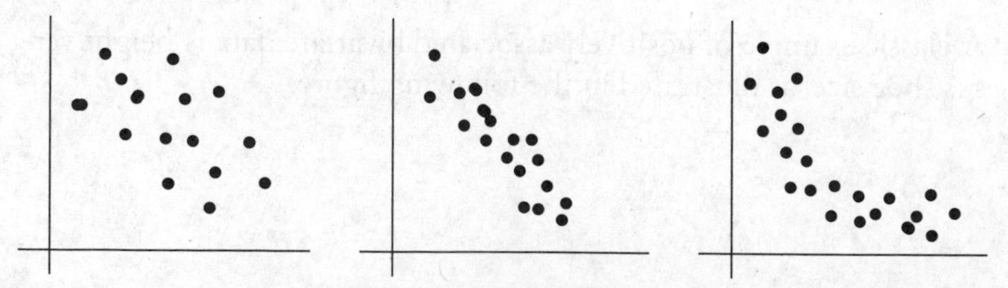

Figure 11–5. Negatively Associated Data

In these three scatterplots, high values of one variable are associated with low values of the other, and vice versa. These are examples of negatively associated data: as one variable goes up, the other goes down. A simple example of negatively associated data is the number of hours spent on work versus the number of hours spent on leisure activities.

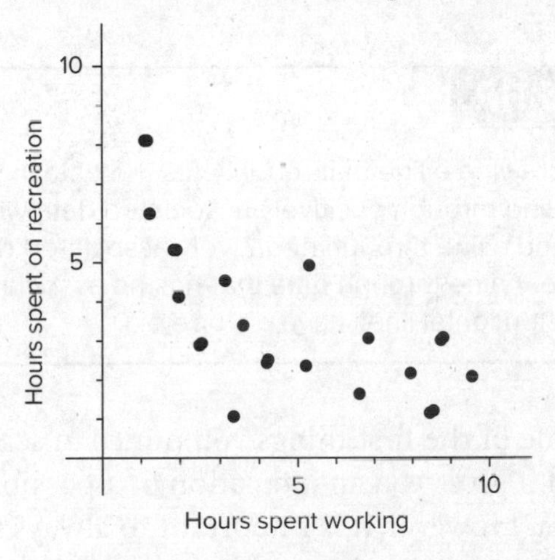

Figure 11–6. Work vs. Play

The data trends downward from left to right, showing a negative association between the two variables. This also probably agrees with your intuition since the more time you spend on work, the less time you would have to spend on leisure.

A scatterplot of bivariate data may show no direction, or association, at all. This may be evidence that there is no relationship between the two variables. Here's a sample scatterplot that shows height versus SAT score.

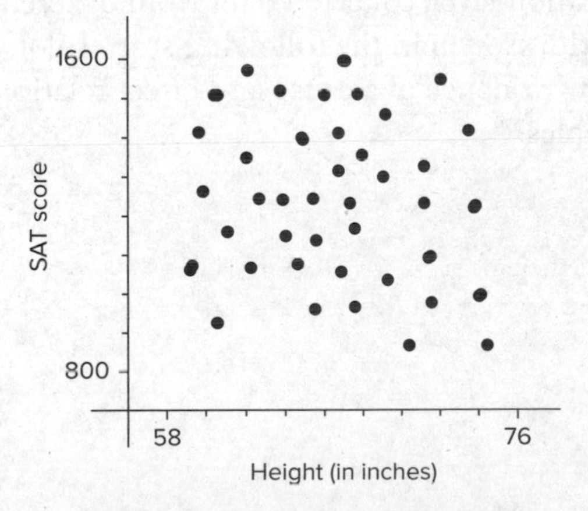

Figure 11–7. Height vs. SAT Score

The data in this scatterplot doesn't appear to be trending up or down. There's no discernible pattern to the data. This makes sense, as there probably isn't any reason why a taller person would score higher, or lower, on the SAT than a shorter person. There's no relationship between these variables.

PAINLESS TIP

If you imagine drawing a line that roughly fits the data in your scatterplot, the line through positively associated data will have a positive slope and a line through negatively associated data will have a negative slope. A line through data that has no association would be horizontal, and horizontal lines have a slope of 0.

Association is one of the first things you notice in scatterplot of bivariate data, and it gives you an indication of a possible relationship between the data. However, it is important to always remember that association does not necessarily point to a cause and effect relationship between the two variables. Just because data values from one variable appear to increase as values from the other increase, that doesn't mean that the increase in one variable is causing the increase in the other. It can feel natural to make this leap in logic, but as you'll see in the next example, you have to be very careful when analyzing bivariate data.

Example 2:

A researcher, who is studying the relationship between life expectancy and education level, collected data from different countries around the world, as seen in the following scatterplot. Does this scatterplot show evidence of a cause and effect relationship between these two variables?

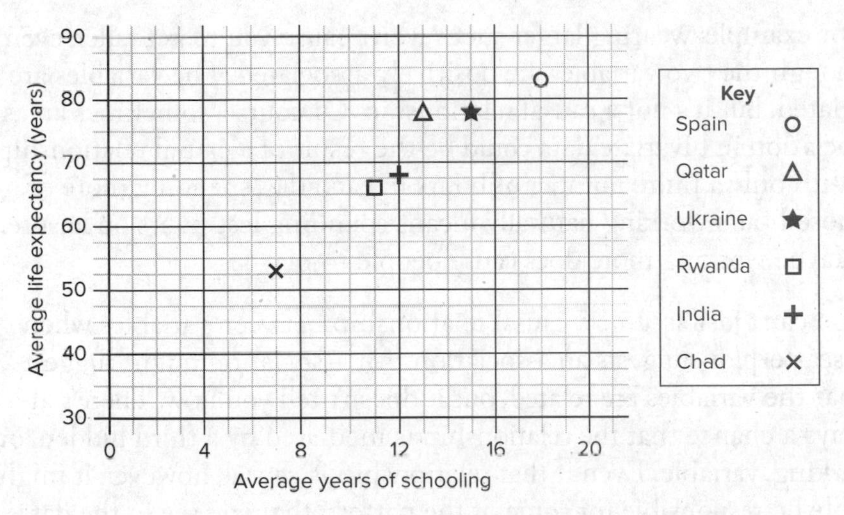

Figure 11–8

There is clearly a positive association in this data. Higher levels of education, measured in years of schooling, are generally paired with higher life expectancies. It's easy to look at this scatterplot and tell a causal story about this association: people who are better educated make better decisions about their health, so they live longer. There may be some truth to this story, but going to school doesn't *cause* you to live longer. It would sound silly to tell someone that attending an extra year of college will add two years to their life span.

The positive association between education and life expectancy doesn't guarantee a causal relationship between the two variables. In fact, there's another explanation for the relationship between these two variables. People in wealthier countries typically tend to live longer and also typically attend more years of school. This is a classic example of a *lurking variable*, a variable unaccounted for in your analysis that potentially impacts the relationship among the variables you are examining. In this scenario, wealth is a lurking variable. Generally speaking, the wealthier a person is, the more schooling that person completes (because, for example, they can afford to delay working). Furthermore, the wealthier a person is, the longer the person lives (because the person can afford better health care). The association you see between education level and life span could really just be the effect of wealth on both of those variables.

In some situations, it's easy to see that a positive association is not the result of a cause and effect relationship between the two variables.

For example, wearing larger shoes won't cause you to get taller, even though the two variables are positively associated. The variables are related, but it's not a causal relationship. Of course, sometimes an association in bivariate data could be the result of a causal relationship. With only a finite number of hours in each day, spending more of those hours working naturally means spending less hours on leisure. Maybe working more does cause people to play less.

You can't just assume a causal relationship between variables when a scatterplot suggests an association. An association might suggest that the variables are related, but it doesn't tell you how. There's always a chance that the relationship is mediated by a third hidden, or lurking, variable. Even if that relationship is causal, however, it might only be responsible for some of the pattern that you see in the data.

CAUTION—Major Mistake Territory!

Association is not causation. People are very good at spotting patterns in data and then constructing stories that explain those relationships, but those stories aren't always true. It takes a lot of analysis and study to really determine the specific nature of the relationship between two sets of data.

Form of Bivariate Data

Positive or negative association is one of the ways a pattern in bivariate data can appear. The form of the data is another. Bivariate data can take different forms, or shapes. In many cases, the scatterplot of bivariate data might look like a blob or a cloud, but sometimes more definite patterns emerge.

Here are some examples of different forms that bivariate data can take.

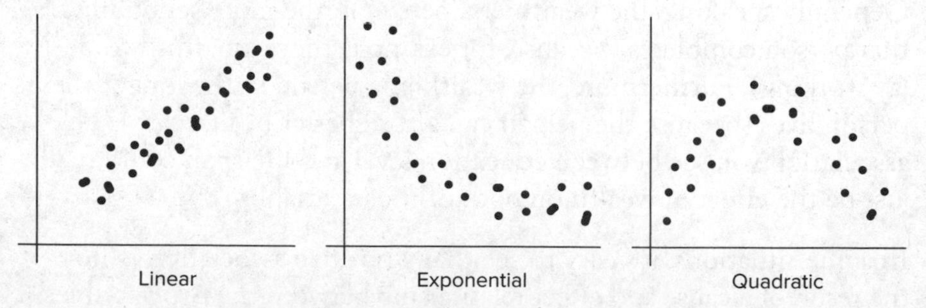

Figure 11–9. Different Forms of Bivariate Data

In the first scatterplot, the data clusters along a line. This is a *linear relationship*, which is one of the most useful relationships bivariate data can have. Much of the remainder of this chapter will deal with quantifying and analyzing linear relationships between the two variables in a bivariate data set.

The second scatterplot shows a negative association, with higher values of the explanatory variable associated with lower values of the response variable, but the form of the data is not linear. Instead, the data clusters around a curve. In this example, the curve is an *exponential* decay curve, which you might see if you study how long it takes patients to fully absorb drug treatments for medical conditions.

In the third scatterplot, there is neither a positive nor a negative association; however, a clear pattern emerges. As you read the scatterplot from left to right, the data rises, peaks, and then falls, which suggests a *quadratic* form of the data, like you would see in a parabola. You might see this pattern in performance data of professional athletes, who first improve with experience but then decline as they age.

Strength of Bivariate Data

The direction and form help you identify the kind of relationship that exists in bivariate data, but the strength of that relationship can be quite different among different variables. Here are three scatterplots that all show positive association and a linear relationship but with different strengths.

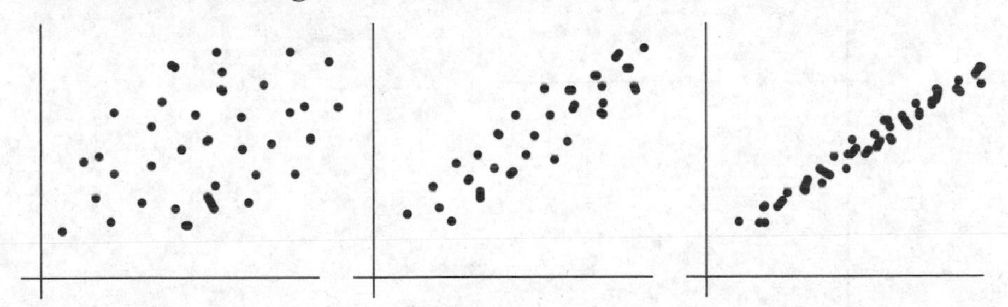

Figure 11–10. Linear Relationships

In the third scatterplot at the far right, the data is very closely clustered around a line. There is a clear linear relationship between the variables, and that relationship is very strong. In the first scatterplot at the far left, there is a general sense of the data moving up and to the right, but the data is less tightly packed around that line. There is still a linear relationship between the variables, but it is a weaker relationship because the data is spread out.

Throughout this book, you have used visualization to help understand single-variable data, and you can do the same with bivariate data using scatterplots. A scatterplot can show you clusters in the data and potential outliers to consider. In addition, it allows you to quickly and easily investigate the relationship in the bivariate data. Your initial observations about the direction and association, form, and strength of that relationship will guide your next steps, which usually involve trying to quantify the nature and strength of the relationship between the two variables. You'll learn about this in the next two sections.

BRAIN TICKLERS Set #33

1. For each of the following scatterplots, indicate whether the scatterplot shows positive association, negative association, or no association.

a.

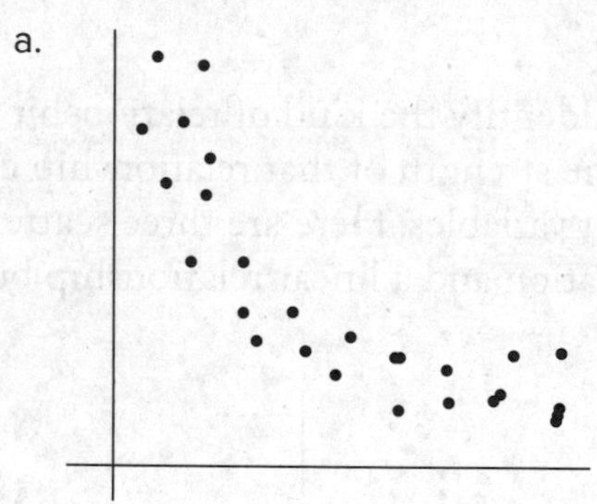

b.

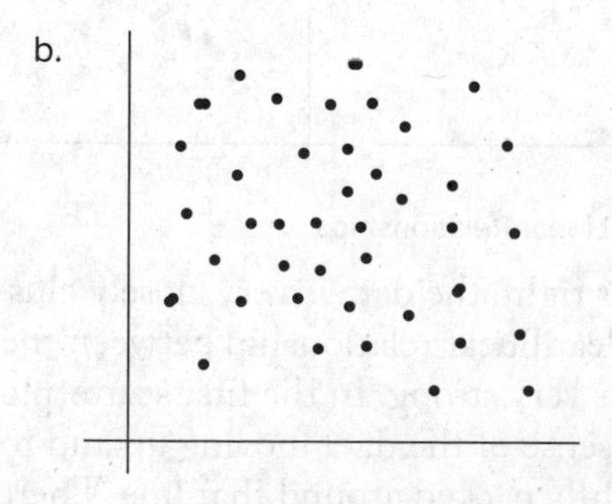

c.

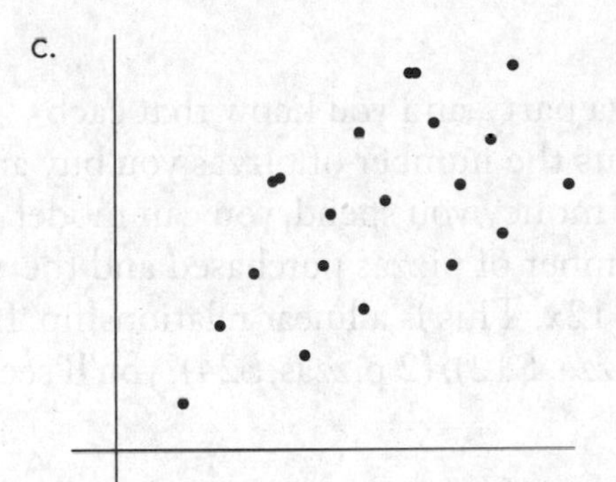

2. What kind of association would you expect between each pair of variables?

   a. Yearly income and amount in savings

   b. Age and the number of books read

   c. Hours spent exercising per week and body mass index (BMI)

3. Sketch a scatterplot of bivariate data that has a strong negative association.

4. Suppose your math test scores are negatively associated with the number of hours of TV you watch per day. What would you expect to happen if you started watching more TV every day?

(Answers are on page 275.)

Correlation

Classifying the direction, form, and strength of a relationship in bivariate data is the first step toward understanding it. Quantifying that relationship is the next step.

The Definition of Correlation

Quantifying relationships between two variables can be tricky business since bivariate data can take many different forms or no apparent form at all. Basic bivariate statistics begins with studying the simplest possible relationship between two variables: a linear relationship. Linear relationships are the easiest kind of relationships to understand and quantify. This kind of relationship appears very frequently among pairs of variables, as seen in the next example.

Example 3:

Suppose you are planning a pizza party, and you know that each pizza pie costs $12. If x represents the number of pizzas you buy and y represents the total amount of money you spend, you can model the relationship between the number of pizzas purchased and the total cost with the equation $y = 12x$. This is a linear relationship. If you plot ordered pairs, like (1 pizza, $12), (2 pizzas, $24), you'll see that the data lies along a line.

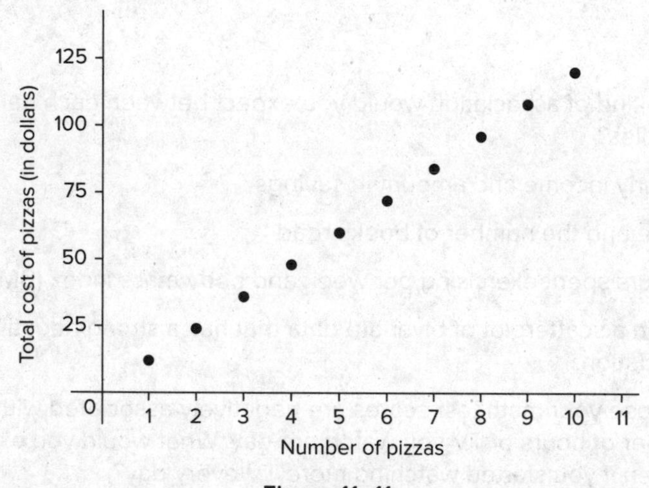

Figure 11–11

In fact, a line passes directly through all of these points because this is a perfect linear relationship.

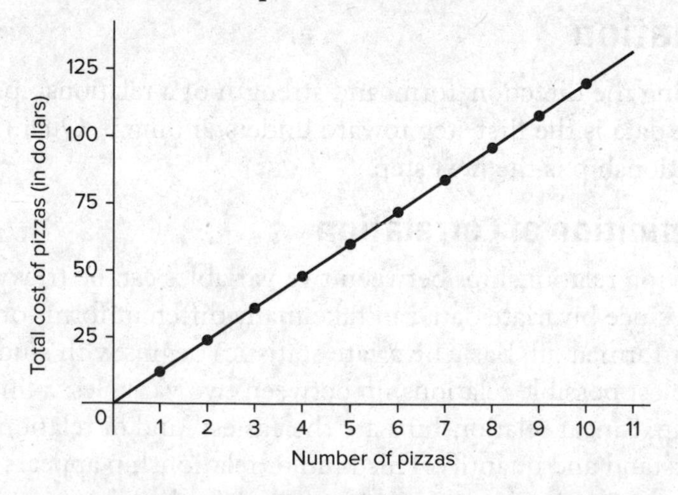

Figure 11–12

If you collected more data from pizza parties around town, you might notice some variation from this line. This variation might be the result of buying pizzas from different restaurants with different prices, different toppings, and different shapes and sizes. That data wouldn't be fit perfectly by a line, but you would still see positive association and a strong linear relationship. The more pizzas you buy, the more money you are going to spend.

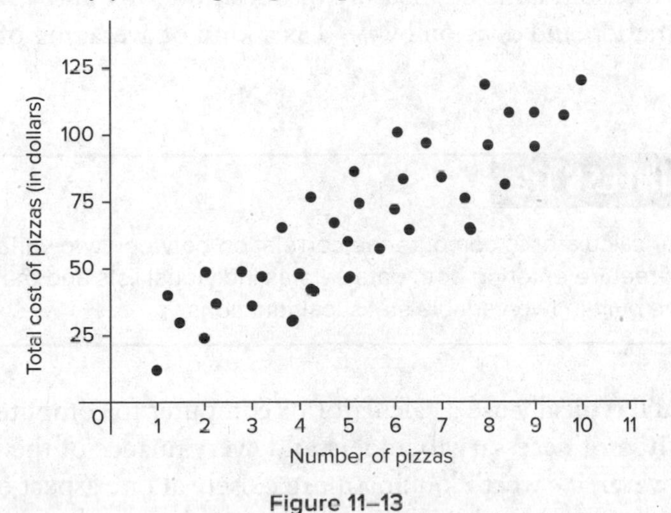

Figure 11–13

Bivariate data will rarely follow a line perfectly, but you can still leverage the linear relationship between the variables for useful analysis. The key is to quantify that relationship. The first step in doing this is through correlation.

Correlation, which is also sometimes known as the *correlation coefficient*, is a way to quantify, or measure, the direction and strength of a linear relationship between two variables. The formula for correlation can be a bit intimidating when you first encounter it, but you've seen everything in the formula before. In fact, you'll probably never calculate a correlation by hand: you'll use calculators or computers to do so.

The Formula for the Correlation Coefficient

Here is the formula that defines the *sample correlation coefficient*, r, between two variables, x and y:

$$r = \frac{1}{n-1}\sum\left(\frac{x-\bar{x}}{S_x}\right)\left(\frac{y-\bar{y}}{S_y}\right)$$

In this formula, $\overline{x}$ and $\overline{y}$ are the means of the two data sets, and S_x and S_y are their sample standard deviations. You've seen the expression $x - \overline{x}$ before: this is just the deviation of x from its mean. Dividing this by S_x gives you $\frac{x - \overline{x}}{S_x}$, which is known as a standard score. This essentially converts each deviation of x into units of standard deviations. For each ordered pair (x, y) in the bivariate data set, you multiply this standardized measure of deviation for x and y, and the summation and division by $n - 1$ is a kind of averaging of those products.

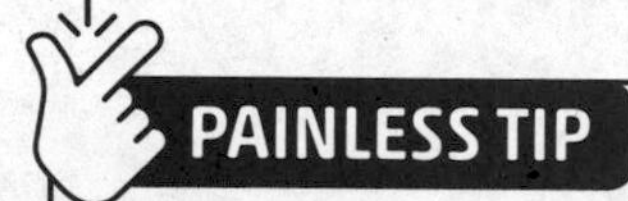

Use your calculator to compute the correlation between two variables. This will require entering both data sets as individual lists and then using the built-in two-variable statistical functions.

Since you'll typically use a calculator or computer to compute correlation, it's not necessary to understand every nuance of the definition. However, it's worth looking more closely at one aspect of the formula. Consider the product of the standard scores inside the summation: $\left(\frac{x - \overline{x}}{S_x}\right)\left(\frac{y - \overline{y}}{S_y}\right)$. First, notice that for each ordered pair (x, y), the size of this product depends on the size of the deviations of x and y from their respective means. The larger those deviations, the larger this product.

Second, this quantity will be positive if x and y are both above or both below their means. In that case, the two deviations will be the same sign—both positive or both negative—and when you multiply them, the result will be positive. This means that the correlation will be positive when x and y are positively associated. In that case, high values of x are paired with high values of y, and low values of x are paired with low values of y, which will produce a positive result in the formula. On the other hand, if high values of x are paired with low values of y and vice versa, that is to say, if the bivariate data is negatively associated, the deviations will have different signs and the product will be negative, producing a negative correlation.

Thus, the sign of the correlation matches the direction, or the association, of the bivariate data. Correlation also captures the strength of the linear relationship. Because of the mathematical properties of the way it is calculated, correlation will always be between –1 and 1. A correlation of 1 is a very strong positive correlation, a correlation of –1 is a very strong negative correlation, and correlations in between represent linear relationships of varying strengths.

> The sample correlation coefficient is sometimes written r_{xy} (read "*r* sub *x y*") to indicate the two variables, *x* and *y*, that are being compared. There is also a *population correlation coefficient*, denoted by ρ, the Greek letter rho, which has a different formula than *r*, but that is beyond the scope of basic bivariate statistics.

Correlation and Association

Here are some scatterplots that show various correlations.

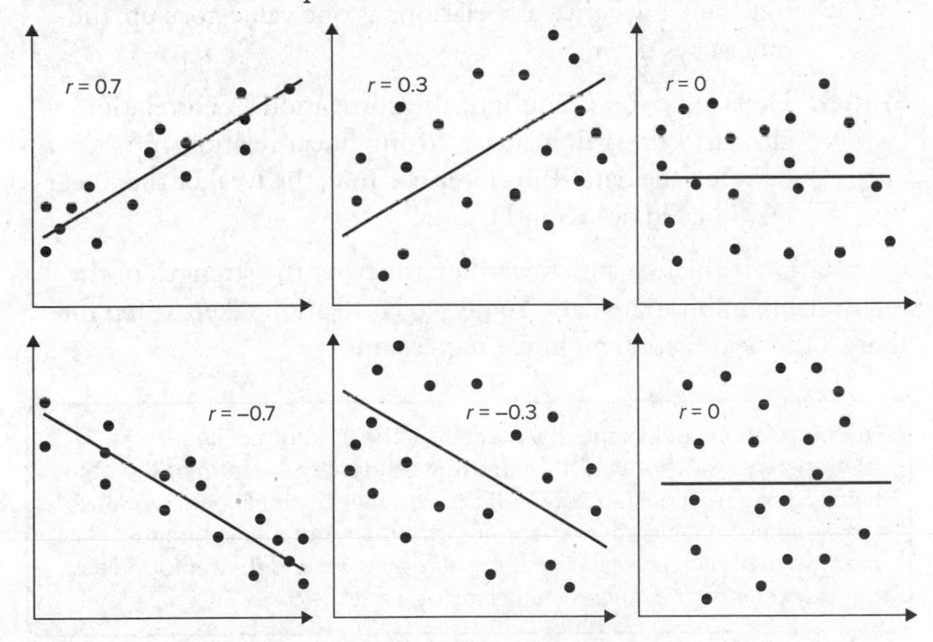

Figure 11–14. Scatterplots and Correlation

Notice that the closer the correlation is to 1, the stronger the positive linear relationship between the variables, and the closer it is to –1, the stronger the negative linear relationship is. When the correlation is 0, you say there is no correlation; this means there is no apparent linear relationship between the variables. Looking back at the formula, you can see that the only way the correlation can end up being 0 is if there are equal measures of pairs of data values that deviate in the same way (that is, on both sides of their respective means) and pairs that deviate in opposite ways (one above and one below the respective mean). This would produce both positive and negative quantities to sum in the calculation of correlation, which would then cancel out each other. In this situation, a large value of x is just as likely to be paired with a large value of y as it is with a small value of y, which tells you there is really no association between the variables.

To interpret correlation, just follow these painless steps.

Step 1: Compute the correlation (r) using a calculator or computer program.

Step 2: Determine whether the correlation is positive or negative. A positive correlation indicates a positive association: as one value goes up, the other goes up. A negative correlation indicates a negative association: as one value goes up, the other goes down.

Step 3: Determine the strength of the correlation. A correlation close to 1 or –1 indicates a strong linear relationship between the data. The closer r is to 0, the weaker the linear relationship between the data.

Correlation is the first step toward quantifying the strength of the relationship in bivariate data. You'll see correlation taken a step further in the next section on linear regression.

> The correlation coefficient only measures the strength of linear relationships, which is why it's important to identify the form of the data initially. If the relationship between the variables is not linear, then this correlation is meaningless. There are other ways of measuring the strength of nonlinear relationships, but they involve different formulas that rely on more advanced statistical ideas.

BRAIN TICKLERS Set #34

1. Which correlation goes with each graph?

i.

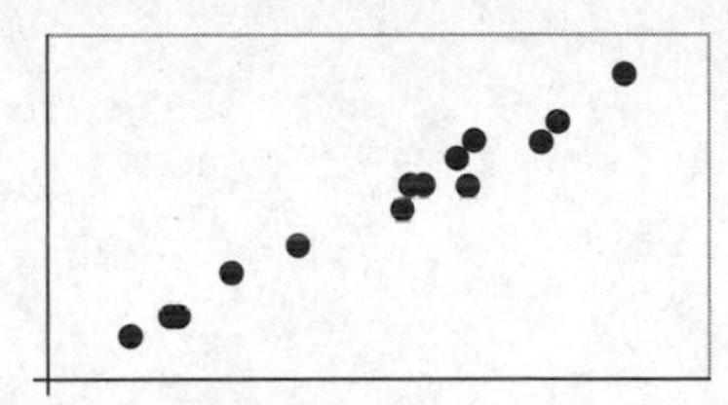

ii.

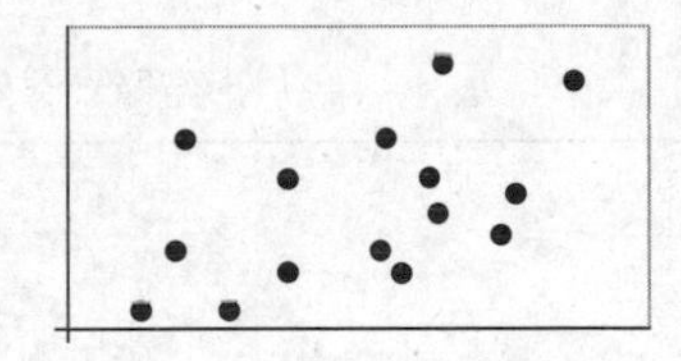

iii.

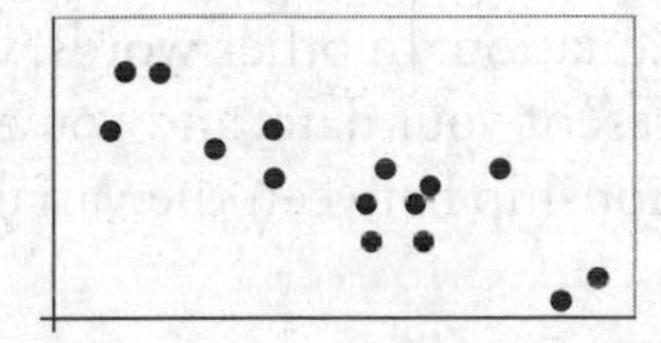

iv.

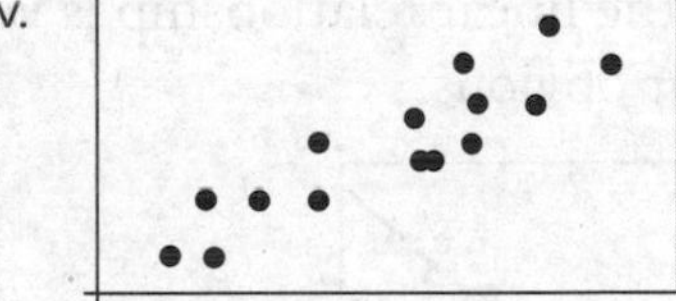

a. $r = 0.75$

b. $r = 0.1$

c. $r = 0.9$

d. $r = -0.65$

2. Would you expect each of the following pairs of data sets to have a strong positive correlation, a weak positive correlation, a strong negative correlation, or a weak negative correlation?

a. Amount of time spent studying and grade point average

b. Age of a chicken and the number of eggs produced

c. Number of hours per day spent playing games and yearly income

3. Use your calculator to compute the sample correlation coefficient of the following data.

x	y
8	10
5	4
12	9
0	2
–6	–2

(Answers are on page 275.)

Linear Regression

Beyond quantifying the direction and strength of a linear relationship, you can model it with a linear equation. In other words, you can actually choose a specific line to represent your data, and you can use that line directly to analyze the relationship between the variables and make projections.

One challenge in this is choosing one line to represent what may appear as a cloud of bivariate data. If the linear relationship is very strong, the choice of line might appear obvious.

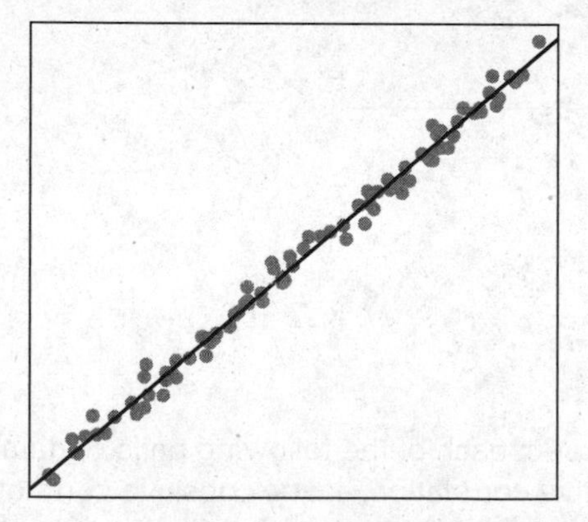

Figure 11–15. Strong Linear Relationship Modeled by Line

However, in a situation where the relationship is linear but weak, many lines might look like they could represent the data.

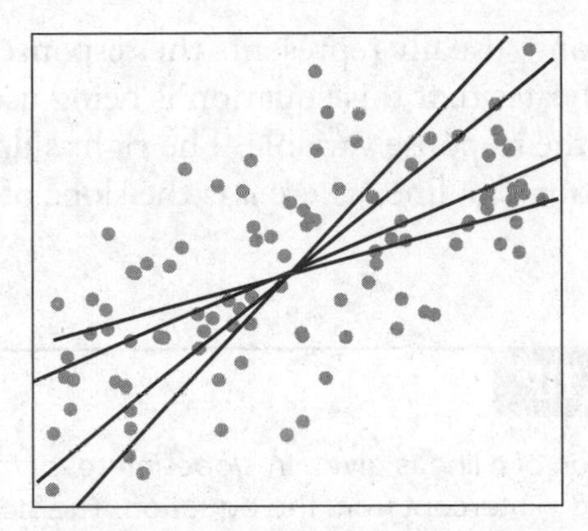

Figure 11–16. Weak Linear Relationship with Multiple Potential Lines

The mathematical solution to this challenge is to come up with a measurement that determines how well a given line fits the data. Once you have a way to measure how closely a line fits the data, you can measure how well every possible line fits your data and then choose the best one. This is called the *line of best fit*, or the *best-fit line*. This is also known as a *regression line*, or *line of regression*.

The Least Squares Regression Line

The most commonly used regression line in statistics is the least squares regression line. With the *least squares regression line*, the so-called *goodness of fit* is measured by computing all the vertical differences between points in the scatterplot and the line, squaring those differences, and then adding them all up. This is a rough measure of the error associated with the line. Different lines will have different errors; the line with the smallest error is the least squares regression line.

A regression line requires an explanatory variable and a response variable, and it gives you an equation that relates them. This is the equation of the line of best fit, the line that best models the linear relationship between the data. In statistics, you will typically use calculators or software packages to find the equations of regression lines.

The equation of a regression line looks just like the regular equation of a line:

$$\hat{y} = ax + b$$

In linear regression, y usually represents the response variable, and $\hat{y}$ (read "y hat") indicates that this equation is being used to estimate or project values of the response variable. The right side of the equation is the usual equation of a line, where a is the slope of the line and b is the y-intercept.

When the equation of a line is given in *slope-intercept form*, you can read its slope and y-intercept from the equation. The slope, which is the constant rate at which the line climbs or falls, is the coefficient of x in the equation. The y-intercept, where the line passes through the y-axis, is the constant term.

The slope of a line tells you how quickly it rises or falls as you read the graph from left to right. A common interpretation of the slope of a line is "rise over run," which is the amount the line changes vertically (the "rise") divided by the amount the line changes horizontally (the "run"). For example, a line with slope 2 goes 2 units up for every 1 unit to the right. A line with slope −0.5 goes half a unit down for every 1 unit it moves to the right.

A large positive slope means the line ascends steeply from left to right, while a large negative slope means the line descends steeply. A slope of zero means the line is horizontal. If this sounds a little like correlation to you, then you won't be surprised to learn that correlation is an important part of the definition of the slope of a regression line.

The slope a of the regression line is given by $r\frac{S_y}{S_x}$. Here, S_x and S_y are the sample standard deviations of the variables x and y, and r is the correlation between them. The formula for the slope offers insight into how the regression line works. If bivariate data is very strongly correlated (that is, if the correlation coefficient is very close to 1), then the slope of the regression line will roughly be equal to $\frac{S_y}{S_x}$. Thinking of this as rise over run, this means that a change of S_x in the x-variable results in a change of S_y in the y-variable. This suggests a strong relationship between the variables: from their respective means, an increase of one standard deviation in the x-variable is associated with an increase of one standard deviation in the y-variable.

The y-intercept of the line, b, can be found by plugging the point $(\bar{x}, \bar{y})$ into the equation of the line. This works because this point, whose x-coordinate is the mean of x and whose y-coordinate is the mean of y, is always on the regression line. This is a consequence of the way the regression line is defined.

Here's an example that demonstrates how to calculate the least squares regression line.

Example 4:

A bivariate data set consists of the amount of time students spent studying for a final exam and their score on that exam. Here is the data.

Time Studying (Hours)	Exam Score
5.5	91
4	93
3	88
4	88
3.5	83
6	99
1.5	95
2	73
2.5	81
7	94

Now, here is a scatterplot of the data.

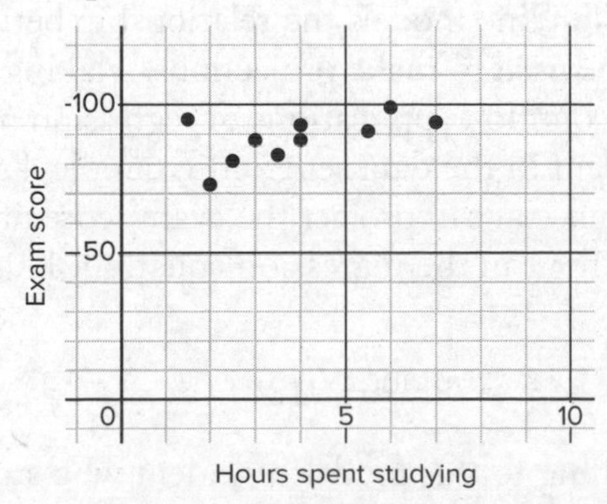

Figure 11–17

You can see that there appears to be a linear relationship between the data and perhaps a slight positive association. The more a student studied, the higher their score on the final exam.

If you enter the data into a calculator or a statistical software package and run a linear regression on the data, you'll find that the equation of the line of regression is $\hat{y} = 2.44x + 78.99$ with a correlation coefficient of $r = 0.57$. Here's the regression line plotted with the data.

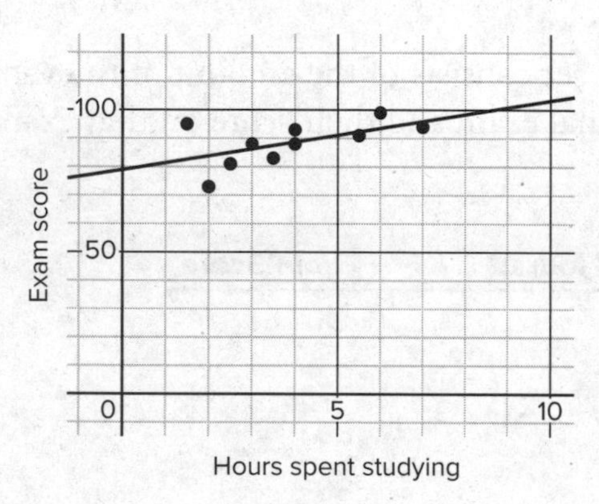

Figure 11–18

You can see that the regression line goes up to the right. This is because the line has a positive slope, which reflects the positive association between the data. Much of the data is close to the line; however, some data points are farther away. There does appear to be a linear relationship, though it isn't especially strong. This is reflected in the correlation coefficient of 0.57.

Making Projections

Since the regression line models the relationship between the two variables, it can be used to make projections. Making projections is one of the most common applications of regression modeling. For example, no student in the data set from Example 4 studied for 5 hours exactly, but you can project the exam score of such a student by plugging in 5 for x in the regression equation. That would give you:

$$\hat{y} = 2.44 \times 5 + 78.99 = 91.2$$

Therefore, according to this model, a student who studies for exactly 5 hours would get a 91 on the exam. This is called interpolation.

Interpolation is when you use a model to project values of the response variable within the range of the explanatory variable.

These kinds of projections are a very useful aspect of regression modeling; however, you have to be careful. If a student studies for this exam for 10 hours, you would compute the student's projected score by substituting 10 in for x in the regression equation:

$$\hat{y} = 2.44 \times 10 + 78.99 = 103.4$$

The model tells you that a student who studies for 10 hours should earn around a 103 on the exam. Assuming that the maximum test score is 100, this projection doesn't make sense. This is an example of extrapolation. *Extrapolation* is when you attempt to project the response variable outside the range of the explanatory variable. In general, extrapolation is less valid than interpolation. A model's projections are most accurate within the initial range of the explanatory variable.

Computing the y-intercept of the regression line is often a kind of extrapolation, one that has a specific interpretation in the context of the linear model. The y-intercept of the regression line is the constant term, but you can think of it as the projection associated with $x = 0$:

$$\hat{y} = 2.44 \times 0 + 78.99 = 78.99$$

This is the projected final exam score of a student who studies for the exam for 0 hours. Whether or not this is a meaningful interpretation depends on the context of the model and also on the extent to which you are extrapolating.

The slope has a particularly useful interpretation in a linear model. Here are two projected exam scores based on three hours of studying ($x = 3$) and four hours of studying ($x = 4$):

$$\hat{y} = 2.44 \times 3 + 78.99 = 86.31$$

$$\hat{y} = 2.44 \times 4 + 78.99 = 88.75$$

Notice that the difference in projected scores is 2.44, which is exactly the slope of the regression line. This is how linear equations work in general. Increasing the input by 1 results in an increase in the output equal to the slope of the line. So according to this model, studying an additional hour for your test will increase your exam score by 2.44 points.

Of course, caution must be applied in drawing such conclusions from linear models. Even though the relationship between these variables appears to be linear, it is not an especially strong relationship since the correlation of 0.57 is not very close to 1. And remember, just because the data shows a positive association doesn't mean there's a causal relationship. As with all statistical processes, there is an inherent amount of uncertainty involved.

Coefficient of Determination

The correlation coefficient can be used to try to quantify some of the uncertainty in a linear relationship. The square of the correlation coefficient, r^2, is called the *coefficient of determination*. This value is interpreted as the percentage of variation in the response variable that is due to variation in the explanatory variable.

In Example 4, $r = 0.57$, so the coefficient of determination is $r^2 = 0.32$. A coefficient of determination of 0.32 suggests that about 32% of the variation in exam scores is due to variation in study time. According to this model, an increase in study time of one hour projects an increase of 2.44 points in the exam score, but because the linear relationship isn't that strong, only around 32% of that increase is due to the increased study time.

Notice that since r is always a number between −1 and 1, the coefficient of determination, r^2, will always be between 0 and 1, so it can always be interpreted as a percentage. In many contexts, the coefficient of determination is a better measure of the strength of a linear correlation than the correlation coefficient itself. An r^2 close to 1 means that the linear relationship between the variables is very strong, while an r^2 close to 0 indicates a very weak linear relationship.

Residuals

Once you have found a linear regression equation, you can further validate your belief that the relationship between the variables is linear by studying the residuals of the data. For each data point in the set, the *residual* is the difference between the actual y-value and the y-value projected by the regression equation. For any data point

(x, y), you simply plug x into the regression equation to compute $\hat{y}$, and the difference, $y - \hat{y}$, is the residual.

Here is the table of data from Example 4, this time with a column added to show the residuals.

Time Studying (Hours)	Exam Score	Residual
5.5	91	−1.41
4	93	4.25
3	88	1.69
4	88	−0.75
3.5	83	−4.53
6	99	5.37
1.5	95	12.35
2	73	−10.87
2.5	81	−4.09
7	94	−2.07

The residuals are essentially the errors between the observed response values and the response values projected by the linear model. It is the sum of the squares of these residuals that is minimized in the least squares regression line. Graphing the residuals in a *residual plot* can help you decide if a linear model is appropriate for your data. Here is the residual plot for this data:

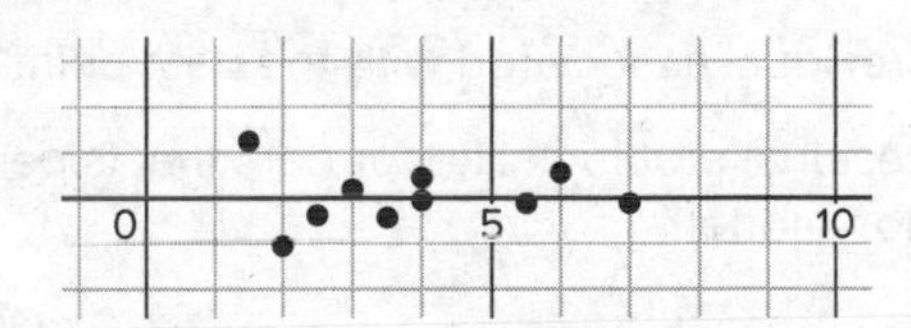

Figure 11–19. Residual Plot for Hours Spent Studying vs. Exam Score

When a residual is close to 0, that means the error in that specific projection is low. When the regression line is above the data in the scatterplot, the residual will be negative, and when the regression line is below the data, the residual will be positive. What you are looking for in a residual plot is a random arrangement of residuals that lie both above and below the zero line. Any pattern you see in the residual plot might indicate an overall pattern in the data that isn't linear, which might cause you to rethink the assumption that the data is linearly related.

> Depending on the form of the relationship in bivariate data, you can find different regression equations. The most common type of regression is linear regression; however, you can find exponential regression equations, trigonometric regression equations, and others.

BRAIN TICKLERS Set #35

1. Use a calculator or statistical software to find the regression line for the following data.

x	y
8	10
5	4
12	9
0	2
−6	−2

2. The least squares regression equation for a set of bivariate data is given to be $\hat{y} = 2.1x + 63.2$.

   a. What is the projection associated with $x = 7$?

   b. What is the residual associated with the data point (10, 81)?

3. What will the residual plot look like for data that is perfectly fit by a linear regression model?

(Answers are on page 276.)

Brain Ticklers—The Answers

Set #33, pages 258–259

1. a. Negative association

   b. No association

   c. Positive association

2. a. Positive

   b. Positive

   c. Negative

3. There are many possible scatterplots that could be drawn. One example is as follows:

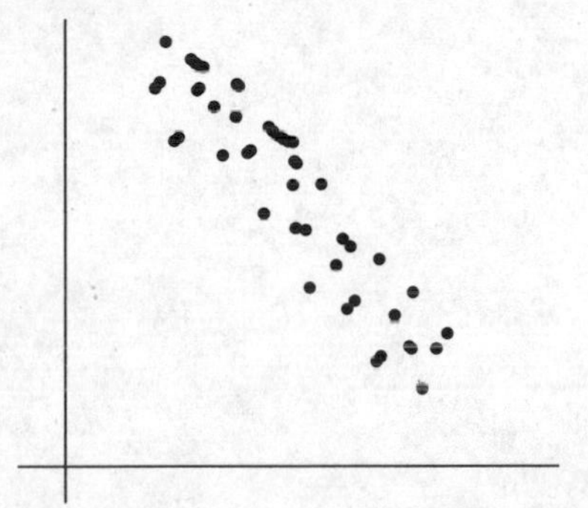

4. Your test scores might go down even further.

Set #34, pages 265–266

1. a. iv

   b. ii

   c. i

   d. iii

2. a. Strong positive correlation

   b. Strong negative correlation

   c. Probably a weak negative correlation

3. $r = 0.949$

Set #35, page 274

1. $\hat{y} = 0.67x + 2.04$
2. a. 77.9

   b. –3.2
3. All the residuals will be zero because there is no error in the model if the line fits the data perfectly.

Chapter 12

Statistical Literacy

Throughout this book, you have learned how to apply the study of statistics to understand and analyze data. However, the most important application of your statistical knowledge may very well be in interpreting the statistical claims of others. Every day, new studies are released, new policies are enacted, and new forecasts are made about the future, all of which rely on data. You can use your knowledge of statistics to put the data, and the claims, in context.

Being statistically literate means understanding enough to know when something is grounded in reasonable thinking and methodology. Statistical literacy is essential to navigating the modern world, and it will only become more important as that world becomes even more data driven. The knowledge you have acquired in this book provides an excellent base for statistical literacy, and in this chapter, you'll learn what to look for as you encounter and interpret the statistics all around you.

Experiments and Studies

Perhaps the most common place you will encounter the application of statistics is in published research. Statistics are used to explore and justify claims made by scientists and doctors but also by companies and political organizations. The goal of statistical research may be to advance scientific knowledge, to motivate a policy shift or corporate strategy, or to establish a connection between cause and effect.

Controlled Experiments

Statistical analysis is frequently part of *controlled experiments*. The goal of an experiment is to impose a *treatment* on a group of *subjects* and observe a response. The treatment is the experimental condition being applied, and the subjects are the people being experimented on.

Experiments are frequently conducted in medical fields but can also be found elsewhere. Here's a simple example of a controlled experiment.

Example 1:

To test a new potential COVID vaccine, a pharmaceutical company might perform an experiment to determine its effectiveness. Subjects would be given the vaccine (the treatment), the results would be observed, and statistics would be used to help put the results in context.

The company might organize an experiment involving 1,000 people. In a *controlled experiment*, one group of subjects would be given the vaccine while another group would be given a placebo. The group given the vaccine would be the *treatment group*, while the group given the placebo would be the *control group*. The response of the treatment group helps the experimenters understand the impact of the treatment, which they can compare to the response of the control group.

> A placebo is essentially a fake treatment. It is provided so that the control group isn't aware that they aren't actually receiving the treatment. This is done because knowing they are, or aren't, receiving treatment can have a psychological impact on the subjects that can influence the results of the experiment.

In this experiment, assuming that the 1,000 subjects are split equally into a treatment group and a control group, the company then observes the subjects to see how many people in each group contract COVID. At the simplest level, if the treatment group contracts COVID at a lower rate than the control group, this might be taken as evidence that the vaccine is effective. The reality is far more complex, but your knowledge of statistics can help the scientists, and you, navigate the situation.

First, suppose that 60 people in the control group contract COVID compared to 30 people in the treatment group contracting COVID.

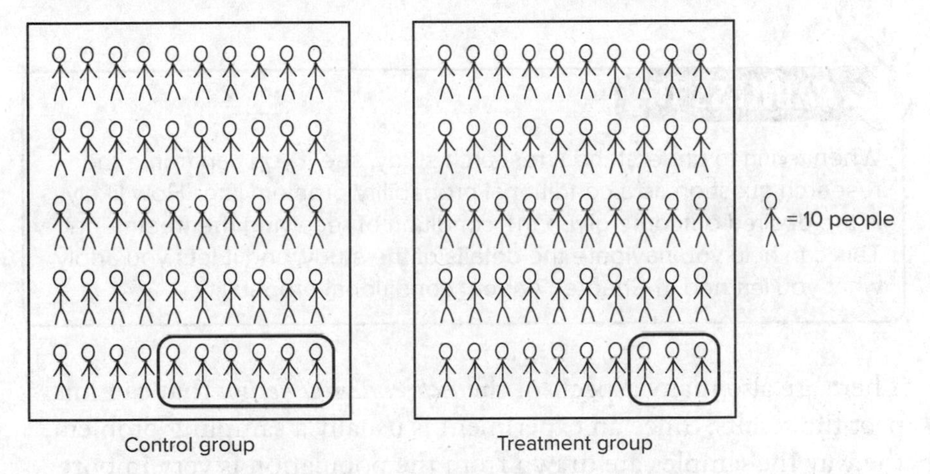

Figure 12–1. Visualization of Subjects Who Contracted COVID

This seems like good evidence for the vaccine's effectiveness, as the rate of infection in the treatment group is half that of the control group. But how strong is that evidence? Is the difference between the treatment and the control significant? This is a situation where hypothesis testing plays an important role.

An experiment is usually a sampling problem, and you know that sampling involves random variation. The results of the experiment suggest that the vaccine has some effect on reducing the prevalence of this illness, but some of the observed difference may be due to random variation. A significance test can help determine how surprised you should be by the results.

Also, notice how this experiment is essentially a conditional probability question. Given the condition of receiving the vaccine, how likely is a subject to contract COVID? Given the condition that the subject received the placebo, what is the probability the subject contracts COVID? And when it comes to determining whether or not the subjects actually have COVID, there is the possibility of both type I and type II errors. A person could test positive for COVID but not actually have it (a type I error), or the person could test negative for COVID even though they do have it (a type II error). All of this must be navigated by the researchers and must be understood by anyone who wants to make sense of the results.

PAINLESS TIP

When trying to understand a research study, see if you can frame the research question as a conditional probability problem like "How likely is the desired outcome given the condition of receiving the treatment?" This can help you navigate the details of the study, and it lets you apply what you learned in Chapter 7 about conditional probability.

There are also many aspects of the *experimental design* that can impact the results. Since an experiment is usually a sampling problem, the way the samples are drawn from the population is very important. Is the sample representative of the population, or were subjects selected in a way that introduced some bias into the experiment? If the vaccine has the potential to work differently on younger people than older people, then a sample skewed toward older people would bias the results. This is an example of why simple random samples are so important in the application of statistical sampling. You have also seen how sample size impacts the variability of sampling and thus the results of the experiment, so this is another factor to consider when designing the experiment.

Not only is overall sample size important, but the sizes of the control and treatment groups can also be relevant. For example, an experiment reporting that only 10 individuals in the control group contracted COVID while 30 individuals in the treatment group contracted COVID might sound like significant evidence against the treatment. However, this result would not be so significant if the control group had only 50 people in it while the treatment group had 500 people.

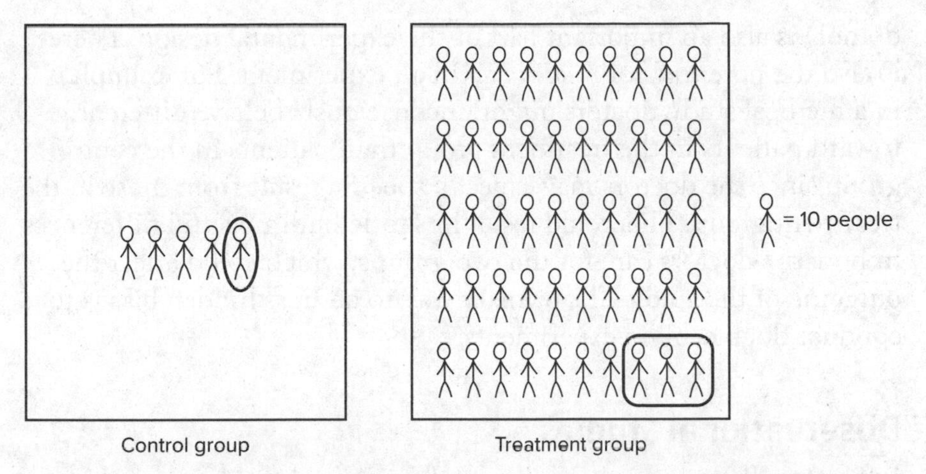

Figure 12–2. Visualization of Subjects Who Contracted COVID

Assuming that the control and treatment groups are comparably sized, assigning a placebo to the control group helps prevent psychological impacts from influencing the results. So, does conducting a *double-blind study*, one in which neither the subjects nor the researchers themselves know who is receiving the treatment. This prevents the researchers from treating the subjects differently (which has the potential to influence the results).

Evaluating Experiments

There are different kinds of *bias* that can interfere with the results of a study. One of the biggest problems with statistical studies is the potential for selection bias. *Selection bias*, also sometimes known as *sampling bias*, occurs when the subjects are chosen in a way that is unrepresentative of the population at large. Improper selection could result in *undercoverage*, when certain subgroups of the population are not properly represented. For example, a study on heart health where the subjects are all men would not be representative of the entire population, so the conclusions would be biased. The potential for selection bias is one of the reasons why random sampling is so important.

In an experiment, it is also important that the treatment and control groups are both representative. Thus, *random assignment* (making sure there is no bias in which subjects get the treatment and which do not) is also an important part of the experimental design. There is also the potential for *hidden bias* in an experiment. For example, in a medical study, doctors might unconsciously behave differently toward patients in the treatment group than patients in the control group since the doctors may expect a specific result from those in the treatment group. This could result in subtle but impactful differences in how the doctors care for the two groups, which could affect the outcome of the study. The primary way to address hidden bias is to conduct double-blind experiments.

Observational Studies

In a controlled experiment, a treatment is given to different subjects, data is collected, and the impact on a response variable is observed. Conducting a controlled experiment, however, isn't always possible. If you are interested in studying existing gender bias in corporate pay, you can't force people to take different kinds of jobs in specific industries. You can only observe the situations that already exist. Nor is a controlled experiment always ethical. If you wanted to study the health effects of using e-cigarettes, it wouldn't be ethical to have people use a product that is assumed to have a negative impact on their health.

In these cases, an observational study is conducted. Unlike in an experiment, in an *observational study*, the researchers don't attempt to influence the results. There is no treatment given to some and not to others. Data is simply observed and analyzed, and attempts are made to find relationships between variables.

There are different kinds of observational studies. In a *longitudinal study*, the same subjects are studied over a period of time and the changes in characteristics are observed. For example, to study the impact of free preschool programs, individuals might be tracked to see how those who participate in free preschool fare later on—in terms of educational attainment, salary, health outcomes, and so on—compared to those who don't attend free preschool.

In a *cross-sectional study*, different people with the same characteristics are compared. Unlike in a longitudinal study, in a cross-sectional study, the data is taken from a single point in time.

Observational studies have the potential to show the same kinds of bias as experiments, like selection bias and hidden bias. So it is important to watch out for this when evaluating the results of studies.

When reviewing a study that involves the application of statistics, you can start to make sense of it by following these painless steps.

Step 1: Identify the type of analysis being done. Is it an experiment or an observational study?

Step 2: Identify the variables being studied and whom the data represents.

Step 3: Examine the experimental design, and be on the lookout for potential selection bias and hidden bias in how the study is conducted.

Here's an example of applying these painless steps to analyze a statistical study like one you might read about.

Example 2:

A company manufactures a pill using natural ingredients that it says improves sleep. To support its claim, the company conducts a study that compares people who use the pill to those who don't. The results of the study show that people who use the pill report sleeping longer and waking up feeling better rested than those who don't. The company then promotes these claims in its advertisements. What might a closer look at the details say about these claims?

Your first step is to try to identify what kind of analysis this is. This kind of information isn't provided in advertisements or even news stories, so you would likely have to read the analysis itself to find out.

Suppose you find the analysis and read about the experimental design. The company selected 50 subjects, gave half of them a free bottle of the pills, and followed up with all 50 subjects two weeks later. So, this is an experiment. Half the group is receiving the treatment (the sleeping aid), and the other half isn't. However, the sample size is small; 50 subjects may not be enough to draw meaningful conclusions about the effectiveness of the sleep aid. Also, this is not a double-blind study. The subjects who get the pill know they are getting the treatment; the subjects who get nothing know they aren't being given the treatment. Furthermore, the researchers also know who gets the treatment and who doesn't.

Now, what variables are being studied? According to the report, subjects who took the pill reported sleeping longer, so one variable of interest is hours of sleep. This is a quantitative variable, but how is it being recorded? Are the subjects being monitored by researchers, or are the subjects recording their own sleep habits? There's a big difference in these two methods. People aren't always objective observers of their own behavior, so if they are self-reporting the information, the data may not be valid.

The subjects also report feeling better rested. This might be a quantitative variable (for example, asking subjects to report how rested they feel on a scale from 1 to 10), but it's highly subjective. One person's "very well rested" may be another person's "slightly rested." This raises questions about how meaningful this data really is.

Lastly, suppose your closer look at the experimental design reveals that the 50 subjects were all customers at a health food store. This is an example of selection bias. Customers from a health food store aren't representative of the population as a whole, so whatever inferences are being drawn are likely unfounded.

Putting this all together, you can see that there are many reasons to doubt this study. People who shop at a health food store are probably more likely to believe a natural sleeping aid could work. Those who are getting the free pills know that they are getting a treatment and may be predisposed to believe that it will have some effect. Thus, when they self-report about their sleep habits, they may be more inclined to believe the pill is having a positive impact on them.

PAINLESS TIP

A healthy skepticism of statistical claims made by advertisers is part of being a good consumer. The more you scrutinize such claims, the more you will see examples of invalid experimental designs, poorly defined variables, and selection bias. As the saying goes, *caveat emptor*: let the buyer beware!

BRAIN TICKLERS Set #36

1. Determine whether each of the following scenarios is more like an experiment or an observational study.

   a. Playing classical music during class to see if there is an impact on exam scores

   b. Analyzing whether people who have pets have better health outcomes

   c. Studying whether offering free breakfast to students has any effects on graduation rates

2. Explain how a study of city traffic that looks at the tolls paid by commuters who cross bridges and tunnels is vulnerable to undercoverage.

3. Explain the potential for hidden bias in a study of educational outcomes where teachers are informed which of their students earned As in their previous class and which earned Cs.

(Answers are on page 296.)

Surveys and Polls

Surveys and polls are common ways to collect data. As you learned earlier in this book, surveys can be an effective way to collect information about a sample, which can be analyzed and used to draw inferences about populations using statistical techniques.

As with experiments and studies, selection bias is a potential problem with surveys. In order for most statistical techniques to be properly applied, samples must be representative of the population, which is true for surveys as well. Polling 1,000 randomly selected Americans and asking their attitudes about religion would produce much different results than if you asked the same question of 1,000 Americans who just got home from church.

A particular kind of selection bias in surveys can result from convenience sampling. Taking random samples of a population can be hard work. *Convenience sampling* occurs when a survey is conducted with people who can be conveniently surveyed. Asking your neighbors or classmates a question is an example of convenience sampling.

You should not expect the results of a convenience sample to be representative of the population.

Survey or polls that you see conducted on social media platforms are usually examples of convenience sampling, as the only people responding are those who are conveniently accessible.

Bias can show up as a result of the way the survey is implemented. For example, any survey conducted online doesn't represent the part of the population that does not have Internet access, so this survey immediately suffers from undercoverage.

A famous example of this kind of bias involves the American presidential election of 1948. An iconic photo shows President-Elect Harry Truman holding up a newspaper with the headline "Dewey Defeats Truman." The newspaper had used a telephone survey to poll Americans about whom they would vote for and determined that Dewey would win comfortably. However, in 1948, the population of Americans who owned phones was not representative of the population overall, which created bias in the survey.

There are other kinds of bias that can arise with surveys. *Voluntary response bias* occurs when respondents are allowed to choose whether or not to participate. People who opt in to participating in a survey—people who respond to an unsolicited email or phone call, or those who are willing to spend 20 minutes participating in market research at the mall—may be a different group of people than the population overall.

The way surveys are designed can also have a big impact on the data that is collected. In particular, the wording and framing of survey questions can affect the responses. Consider the difference between the following questions:

- Should the U.S. government spend less money on national defense?
- Should the U.S. government decrease its $800 billion per year military budget?

The sentiment behind these questions is roughly equivalent, but the different wordings will likely produce different responses. People might be more reluctant to say yes to the first question, because a country needs to be able to defend itself. However, people may be more prone to say yes to the second question, because $800 billion is a big number and many people assume that governments spend money inefficiently. Designing survey questions that avoid this potential bias requires knowledge of surveying techniques and of psychology. But, in the wrong hands, knowledge of how people respond to different types of survey questions can be used to design surveys that produce desired outcomes, like support for a specific product or policy.

When examining the results of a survey or poll, here are some painless steps to make sure the results won't lead you astray.

Step 1: Determine whether or not the sample is representative of the population.

Step 2: Look for potential response bias in the survey design.

Step 3: Examine the wording of the questions to see if that might have impacted how people responded.

Here's an example of evaluating a survey using these painless steps.

Example 3:

A large city is considering a congestion pricing plan that will levy a toll on cars that drive into the downtown area. The purpose of the plan is to reduce automobile traffic in busy areas while raising funds for environmental cleanup. An advocacy group conducts a survey about how residents feel about the plan. Here's a pie chart that shows the results of that survey.

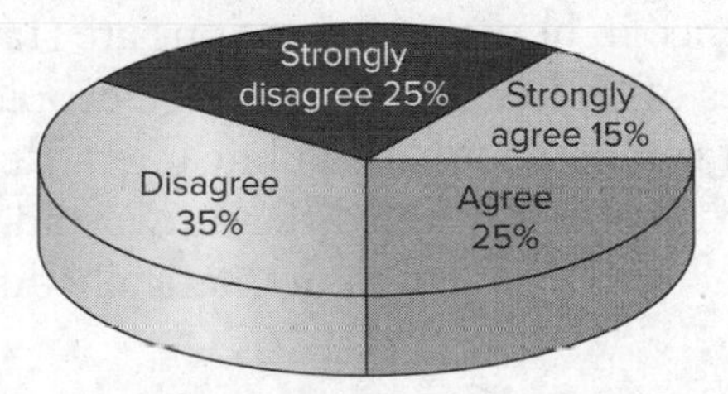

Figure 12–3. Public Opinion About Congestion Pricing Plan

This pie chart suggests that the public is not very supportive of this plan, with 60% saying they disagree or strongly disagree. However, a closer look might change your interpretation of the results.

The first question you should ask is "Who was surveyed?" Suppose you read the fine print and discover that the advocacy group got a list of names and addresses from the Department of Motor Vehicles. The group mailed out a survey about the congestion pricing plan to 500 automobile owners, and 50 responded. Immediately, you should recognize the potential for selection bias. Not everyone owns a car, and people who own cars will be affected differently by a plan that imposes a surcharge on driving than those who don't own a car. Therefore, this sample is not representative of the population.

There's also the potential for response bias. Of the 500 people who received the survey, only 50 people responded. People who feel very strongly about an issue are probably more likely to take the time to fill out and return a survey, which could skew the results.

Then, there's the issue of what the survey actually asked. Suppose this was the question respondents answered:

> "How do you feel about the city's proposed congestion pricing plan that will cost the average driver an additional $250 per year in taxes?"

Now, imagine this alternative wording:

> "How do you feel about the city's plan to collect a $1 per day surcharge on drivers entering the downtown area to reduce traffic and combat climate change?"

Notice the difference in how these questions are framed. The first evokes images of higher taxes on average people; the second suggests a modest surcharge for the public good. Respondents will have different reactions to the different wordings. As a result, those different reactions can lead to dramatically different survey results.

CAUTION—Major Mistake Territory!

Not all surveys are created equal. A scientific survey is likely to have been designed to counteract potential selection bias and response bias. A survey from a corporation or advocacy group might very well be designed to capitalize on selection bias and response bias. Don't assume a survey was designed to eliminate bias. In fact, it might have been designed to capitalize on bias! Always check the source of the survey, examine how the survey was conducted, and pay attention to the details.

BRAIN TICKLERS Set #37

1. Explain how the following scenarios could potentially involve selection bias.
   a. Polling 100 people at a museum, asking their opinion about government funding for the arts
   b. Polling 100 people who are attending a high school football game, asking their opinion about raising taxes to provide more funds for education
   c. Polling 100 students, asking their opinion about whether the school year should be extended from 10 months to 12 months
2. What's the problem with mailing out a survey to random people and asking them to mail their answers back to the questioner?
3. What's the potential problem with this survey question: "How frustrated are you with the government's handling of the current economic crisis?" How might it be better worded?

(Answers are on page 296.)

Evaluating the Validity of Statistical Arguments

News stories are often full of data: economic indicators, educational outcomes, climate statistics, and much more. These news stories are frequently reports of studies that rely on statistics. The knowledge you have developed throughout this book and in this chapter, however, can help you make sense of what you hear on television, in the newspapers, and on social media.

In this chapter, you've learned about the different kinds of studies that use statistics and what to watch out for when analyzing and interpreting those studies. Whether it's an experiment, observational study, or survey, the design is crucial in producing valid data. Has care been taken to avoid selection bias in the way samples are produced? Are the underlying processes—like sample selection and assignment of treatment—random enough to satisfy the requirements of statistical sampling? There are also many other details to consider when evaluating the validity of statistical arguments. Here are a few.

Proxy Measurement

Measurement is often at the heart of scientific and statistical study. Some characteristics, like income, height, or blood pressure, are easy to measure. Other quantities, like quality of life, customer satisfaction, and happiness, are much more difficult to measure. In cases like this, proxy measurements are often used.

A *proxy* is a stand-in for something, so a *proxy measurement* is the measurement of one quantity to stand in for another. For example, a company might measure customer satisfaction by looking at the frequency of repeat business. In this case, additional future purchases might be considered as a proxy for how satisfied a customer is with a company's products and services.

Proxy measurement is useful but can also be misleading. It is often assumed that intelligence is measured by an IQ test, but human intelligence is far more complex and wide-ranging than the results of an IQ test could capture. The IQ test is a proxy measurement. Similarly, success in school is often measured as a grade point average, but this too is an imperfect proxy measurement. Success in school could take many forms outside of high course grades (for example, success in sports, leadership in student clubs, participation in vocational programs, achievement in the arts, and so on).

CAUTION—Major Mistake Territory!

Measurements in statistical studies are often proxy measurements. Be careful not to confuse the proxy measurement with the actual quantity you are interested in understanding.

Replication and *p*-Hacking

The significance tests covered in Chapter 10 are used in many scientific and medical studies and are commonly reported on in popular news media. However, such reports should be read cautiously. First, you know that many different factors—such as the sample size and properties of estimators—must be satisfied in order to proceed with significance tests. You also know that the word "significant" has a very technical meaning in this context. Statistical significance is a specific claim about properties of samples in the sampling distribution. A casual reader will probably just think that "significant" means important. However, you know this level of importance is a conventional decision.

A 5% significance level is the standard in many fields. So achieving a result at this level of significance can mean the difference between appearing to discover something new and having the experiment be considered a failure. This can result in *p-hacking*, the process of redesigning and rerunning an experiment until a desirable level of significance has been achieved.

More than that, you also know that a result that is significant at the 5% level will occur 5% of the time just due to random chance. That is to say, there is a 5% chance of a type I error. So, even when there is no phenomenon to actually be observed, one in every 20 studies might show evidence of the phenomenon. This is part of the *replication crisis* facing some scientific research. Some established studies have failed to be *replicated*, meaning that other teams of researchers have attempted to reproduce the experiment but did not get the same results as the original researchers. This could possibly be because the original research was actually a result of random variation rather than of the proposed phenomenon.

As you learned in Chapter 10, the word "significant" has a technical meaning in statistics regarding the likelihood of observing a specific type of result in a sampling process. When reading about research, don't automatically interpret "significant" as "important" or "meaningful."

Correlation, Causation, and Confounding

As you learned in Chapter 11, there is a lot to be cautious about when it comes to bivariate data. "Correlation is not causation" is one of the most common phrases you'll hear in popular discussions of statistics; it refers to what you learned about association. Just because high values of one variable are associated with high values of the other doesn't mean that one high value causes the other.

One reason for this is the possibility of lurking variables. As you learned in Chapter 11, a lurking variable is a hidden variable that may explain the relationship between the variables being studied. Education and life span may be positively associated, but it's the relationship that wealth has on both that explains their relationship.

Confounding is another issue that complicates drawing conclusions from bivariate data. *Confounding* occurs when the effect of the explanatory variable on the response variable is inextricably linked to one or more other variables that impact the response variable. For example, if a study shows that students who take Advanced Placement (AP) courses in high school graduate college at higher rates, that might encourage school districts to offer AP courses at more schools. Yet whether or not schools already offer AP courses and prepare students to succeed in them is intricately connected to a variety of other school quality factors—such as per pupil spending, class size, and teacher qualifications—all of which confound the relationship between AP scores and college graduation rates. In these situations, researchers often try to isolate or *control* for those confounding factors, but in general, this is very difficult to do.

One of the most common statistical applications in the analysis of bivariate data is regression, which fits a mathematical model to the data. Linear regression starts with the assumption that the relationship between the data is linear. Thus, there is always the danger of seeing evidence of a linear relationship even though there isn't one. There are also dangers associated with *extrapolation*, which means using a regression equation to project values outside the original range of the data. For example, if you look at world records in the 100-meter dash, you can fit a linear model to the data and see a negative trend: world record times are decreasing every year. But, if you extrapolate this to the extreme, you'll see your model eventually predicting a human being running the race in under 1 second.

Just because bivariate data shows a linear relationship in a scatterplot doesn't mean there is a cause and effect relationship between the variables.

Other Types of Bias

There are many types of bias to watch out for in studies and surveys. You've already learned about selection bias and hidden bias. Here are a couple of others.

Confirmation bias occurs when you collect data only from those who are likely to confirm your hypothesis. For example, a pizza company asking its customers if they prefer pizza to burgers would be an example of confirmation bias: people who buy pizza are probably more likely to say they prefer pizza to burgers than a random sample of the population. A musician asking fans for feedback on their latest album is also an example of confirmation bias: someone who is already a fan is probably predisposed to like that artist's work.

Survivorship bias occurs when you only collect data from those individuals who remain at the end of some process. For example, asking customers to rate their satisfaction with their customer service experience at the end of a long call may suffer from survivorship bias, as many of the most frustrated customers might have hung up rather than wait. Likewise, asking recent graduates if their university was a good choice is also vulnerable to survivorship bias, as it ignores those students who left the university before graduating.

Statistics in the News

There is no shortage of statistical stories in the news, but it's important to remember that reporting about a statistical study is something different than the study itself. Journalists may focus on the boldest claims of a scientific result and ignore the many qualifying, but relevant, details that are part of the study. They may also simply misinterpret or misunderstand what a study says. That's why it's important for you to develop your own sense of statistical literacy to judge for yourself. A good step is to look beyond the story to the study itself, as seen in the following example.

Example 4:

A common story in educational research is about how certain factors—a superior teacher, a charter school, an educational intervention—can lead to increased learning for students. Such research has great potential value in improving education, but it's important to go beyond the headlines and pay attention to the details of the studies themselves.

A story covering one particular school district might claim that students in that district learn more than students in another district, citing a study that measures years of learning for students in comparable grades.

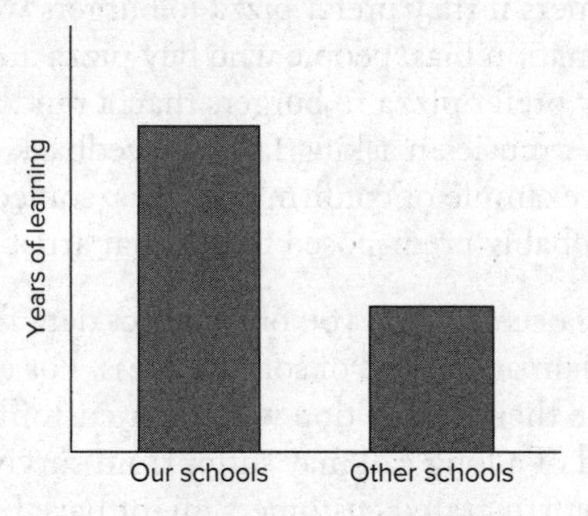

Figure 12–4

Such an impressive and important-sounding claim deserves attention, but like all statistical reporting, it deserves thoughtful scrutiny as well. Furthermore, when you start asking the questions that have been outlined in this chapter and in this book—questions about experimental design, the definition of variables, and what is really being measured—a very different story might emerge.

It might turn out that the study only considered a very small number of students in a particular educational setting, so the sample size might be an issue. It might be that this particular school district serves a different population of students than the district it's being compared with, which would be an issue of selection bias. And then there's the question of measurement. What does it mean that one student has learned 6 months more material than another? Often, what's really being measured is the result of a test. Getting one or two more questions right on that test is being interpreted as 6 or 8 months of additional learning. This is an issue of proxy measurement and interpretation.

Data and statistics play an important role in our lives and in our world, and so they will continue to be a part of the stories that we tell and the decisions that we make. Use the knowledge you have gained in this book to help put those stories, and those statistics, in context. Hopefully what you've learned can help make navigating our complex, data-driven world a little more painless.

BRAIN TICKLERS Set #38

1. What are some proxy measurements for school quality?
2. A political figure asks their followers on social media whether they agree with the politician's position on an issue. What kind of bias is this an example of?
3. Smartphone price data from the past 10 years is collected and analyzed, and a linear regression model predicts that 20 years from now, top-quality smartphones will cost less than $100. What's the problem with this conclusion?

(Answers are on page 296.)

Brain Ticklers—The Answers

Set #36, page 285

1. a. Experiment

   b. Observational study

   c. Observational study
2. Since only car drivers pay such tolls, the study would miss the large segment of the population that uses public transportation to commute to work and school.
3. Teachers might treat students differently based on knowledge of their past performance, which could influence the study.

Set #37, page 289

1. a. People at a museum probably have a much more favorable opinion about the arts.

   b. Attendees of a high school football game are more likely to be directly affected by school quality, so they will likely have more favorable attitudes about funding schools.

   c. Students would lose vacation under a longer school year and might be more likely to oppose it.
2. This is an example of voluntary response bias. The surveys were distributed randomly, but those who actually go to the trouble to mail their responses back might not be representative of the population.
3. This question plants the idea of crisis and frustration in the respondent's mind. A more neutral phrasing might be "What is your opinion about how the government is managing the economy?"

Set #38, page 295

1. There are many possibilities. Examples include graduation rates, standardized test scores, postschool employment rates, how long the wait list is to get into the school, self-reported satisfaction scores, and so on.
2. Confirmation bias. It is also an example of convenience sampling.
3. The model is being extrapolated well beyond the range of the original data to the point where it may be meaningless.

Appendix: Standard Normal Table

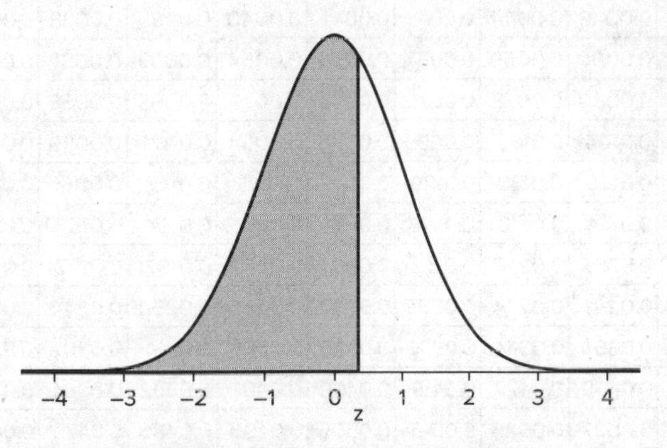

Figure A–1. Table entry for *z* is the area under the curve to the left of *z*

z	0.00	0.01	0.02	0.03	0.04	0.05	0.06	0.07	0.08	0.09
–3.5	0.0002	0.0002	0.0002	0.0002	0.0002	0.0002	0.0002	0.0002	0.0002	0.0002
–3.4	0.0003	0.0003	0.0003	0.0003	0.0003	0.0003	0.0003	0.0003	0.0003	0.0002
–3.3	0.0005	0.0005	0.0005	0.0004	0.0004	0.0004	0.0004	0.0004	0.0004	0.0003
–3.2	0.0007	0.0007	0.0006	0.0006	0.0006	0.0006	0.0006	0.0005	0.0005	0.0005
–3.1	0.0010	0.0009	0.0009	0.0009	0.0008	0.0008	0.0008	0.0008	0.0007	0.0007
–3.0	0.0013	0.0013	0.0013	0.0012	0.0012	0.0011	0.0011	0.0011	0.0010	0.0010
–2.9	0.0019	0.0018	0.0018	0.0017	0.0016	0.0016	0.0015	0.0015	0.0014	0.0014
–2.8	0.0026	0.0025	0.0024	0.0023	0.0023	0.0022	0.0021	0.0021	0.0020	0.0019
–2.7	0.0035	0.0034	0.0033	0.0032	0.0031	0.0030	0.0029	0.0028	0.0027	0.0026
–2.6	0.0047	0.0045	0.0044	0.0043	0.0041	0.0040	0.0039	0.0038	0.0037	0.0036
–2.5	0.0062	0.0060	0.0059	0.0057	0.0055	0.0054	0.0052	0.0051	0.0049	0.0048
–2.4	0.0082	0.0080	0.0078	0.0075	0.0073	0.0071	0.0069	0.0068	0.0066	0.0064
–2.3	0.0107	0.0104	0.0102	0.0099	0.0096	0.0094	0.0091	0.0089	0.0087	0.0084
–2.2	0.0139	0.0136	0.0132	0.0129	0.0125	0.0122	0.0119	0.0116	0.0113	0.0110
–2.1	0.0179	0.0174	0.0170	0.0166	0.0162	0.0158	0.0154	0.0150	0.0146	0.0143
–2.0	0.0228	0.0222	0.0217	0.0212	0.0207	0.0202	0.0197	0.0192	0.0188	0.0183
–1.9	0.0287	0.0281	0.0274	0.0268	0.0262	0.0256	0.0250	0.0244	0.0239	0.0233
–1.8	0.0359	0.0351	0.0344	0.0336	0.0329	0.0322	0.0314	0.0307	0.0301	0.0294
–1.7	0.0446	0.0436	0.0427	0.0418	0.0409	0.0401	0.0392	0.0384	0.0375	0.0367
–1.6	0.0548	0.0537	0.0526	0.0516	0.0505	0.0495	0.0485	0.0475	0.0465	0.0455
–1.5	0.0668	0.0655	0.0643	0.0630	0.0618	0.0606	0.0594	0.0582	0.0571	0.0559
–1.4	0.0808	0.0793	0.0778	0.0764	0.0749	0.0735	0.0721	0.0708	0.0694	0.0681
–1.3	0.0968	0.0951	0.0934	0.0918	0.0901	0.0885	0.0869	0.0853	0.0838	0.0823
–1.2	0.1151	0.1131	0.1112	0.1093	0.1075	0.1056	0.1038	0.1020	0.1003	0.0985
–1.1	0.1357	0.1335	0.1314	0.1292	0.1271	0.1251	0.1230	0.1210	0.1190	0.1170
–1.0	0.1587	0.1562	0.1539	0.1515	0.1492	0.1469	0.1446	0.1423	0.1401	0.1379
–0.9	0.1841	0.1814	0.1788	0.1762	0.1736	0.1711	0.1685	0.1660	0.1635	0.1611
–0.8	0.2119	0.2090	0.2061	0.2033	0.2005	0.1977	0.1949	0.1922	0.1894	0.1867
–0.7	0.2420	0.2389	0.2358	0.2327	0.2296	0.2266	0.2236	0.2206	0.2177	0.2148
–0.6	0.2743	0.2709	0.2676	0.2643	0.2611	0.2578	0.2546	0.2514	0.2483	0.2451
–0.5	0.3085	0.3050	0.3015	0.2981	0.2946	0.2912	0.2877	0.2843	0.2810	0.2776
–0.4	0.3446	0.3409	0.3372	0.3336	0.3300	0.3264	0.3228	0.3192	0.3156	0.3121
–0.3	0.3821	0.3783	0.3745	0.3707	0.3669	0.3632	0.3594	0.3557	0.3520	0.3483
–0.2	0.4207	0.4168	0.4129	0.4090	0.4052	0.4013	0.3974	0.3936	0.3897	0.3859
–0.1	0.4602	0.4562	0.4522	0.4483	0.4443	0.4404	0.4364	0.4325	0.4286	0.4247
–0.0	0.5000	0.4960	0.4920	0.4880	0.4840	0.4801	0.4761	0.4721	0.4681	0.4641

z	**0.00**	**0.01**	**0.02**	**0.03**	**0.04**	**0.05**	**0.06**	**0.07**	**0.08**	**0.09**
0.0	0.5000	0.5040	0.5080	0.5120	0.5160	0.5199	0.5239	0.5279	0.5319	0.5359
0.1	0.5398	0.5438	0.5478	0.5517	0.5557	0.5596	0.5636	0.5675	0.5714	0.5753
0.2	0.5793	0.5832	0.5871	0.5910	0.5948	0.5987	0.6026	0.6064	0.6103	0.6141
0.3	0.6179	0.6217	0.6255	0.6293	0.6331	0.6368	0.6406	0.6443	0.6480	0.6517
0.4	0.6554	0.6591	0.6628	0.6664	0.6700	0.6736	0.6772	0.6808	0.6844	0.6879
0.5	0.6915	0.6950	0.6985	0.7019	0.7054	0.7088	0.7123	0.7157	0.7190	0.7224
0.6	0.7257	0.7291	0.7324	0.7357	0.7389	0.7422	0.7454	0.7486	0.7517	0.7549
0.7	0.7580	0.7611	0.7642	0.7673	0.7704	0.7734	0.7764	0.7794	0.7823	0.7852
0.8	0.7881	0.7910	0.7939	0.7967	0.7995	0.8023	0.8051	0.8078	0.8106	0.8133
0.9	0.8159	0.8186	0.8212	0.8238	0.8264	0.8289	0.8315	0.8340	0.8365	0.8389
1.0	0.8413	0.8438	0.8461	0.8485	0.8508	0.8531	0.8554	0.8577	0.8599	0.8621
1.1	0.8643	0.8665	0.8686	0.8708	0.8729	0.8749	0.8770	0.8790	0.8810	0.8830
1.2	0.8849	0.8869	0.8888	0.8907	0.8925	0.8944	0.8962	0.8980	0.8997	0.9015
1.3	0.9032	0.9049	0.9066	0.9082	0.9099	0.9115	0.9131	0.9147	0.9162	0.9177
1.4	0.9192	0.9207	0.9222	0.9236	0.9251	0.9265	0.9279	0.9292	0.9306	0.9319
1.5	0.9332	0.9345	0.9357	0.9370	0.9382	0.9394	0.9406	0.9418	0.9429	0.9441
1.6	0.9452	0.9463	0.9474	0.9484	0.9495	0.9505	0.9515	0.9525	0.9535	0.9545
1.7	0.9554	0.9564	0.9573	0.9582	0.9591	0.9599	0.9608	0.9616	0.9625	0.9633
1.8	0.9641	0.9649	0.9656	0.9664	0.9671	0.9678	0.9686	0.9693	0.9699	0.9706
1.9	0.9713	0.9719	0.9726	0.9732	0.9738	0.9744	0.9750	0.9756	0.9761	0.9767
2.0	0.9772	0.9778	0.9783	0.9788	0.9793	0.9798	0.9803	0.9808	0.9812	0.9817
2.1	0.9821	0.9826	0.9830	0.9834	0.9838	0.9842	0.9846	0.9850	0.9854	0.9857
2.2	0.9861	0.9864	0.9868	0.9871	0.9875	0.9878	0.9881	0.9884	0.9887	0.9890
2.3	0.9893	0.9896	0.9898	0.9901	0.9904	0.9906	0.9909	0.9911	0.9913	0.9916
2.4	0.9918	0.9920	0.9922	0.9925	0.9927	0.9929	0.9931	0.9932	0.9934	0.9936
2.5	0.9938	0.9940	0.9941	0.9943	0.9945	0.9946	0.9948	0.9949	0.9951	0.9952
2.6	0.9953	0.9955	0.9956	0.9957	0.9959	0.9960	0.9961	0.9962	0.9963	0.9964
2.7	0.9965	0.9966	0.9967	0.9968	0.9969	0.9970	0.9971	0.9972	0.9973	0.9974
2.8	0.9974	0.9975	0.9976	0.9977	0.9977	0.9978	0.9979	0.9979	0.9980	0.9981
2.9	0.9981	0.9982	0.9982	0.9983	0.9984	0.9984	0.9985	0.9985	0.9986	0.9986
3.0	0.9987	0.9987	0.9987	0.9988	0.9988	0.9989	0.9989	0.9989	0.9990	0.9990
3.1	0.9990	0.9991	0.9991	0.9991	0.9992	0.9992	0.9992	0.9992	0.9993	0.9993
3.2	0.9993	0.9993	0.9994	0.9994	0.9994	0.9994	0.9994	0.9995	0.9995	0.9995
3.3	0.9995	0.9995	0.9995	0.9996	0.9996	0.9996	0.9996	0.9996	0.9996	0.9997
3.4	0.9997	0.9997	0.9997	0.9997	0.9997	0.9997	0.9997	0.9997	0.9997	0.9998
3.5	0.9998	0.9998	0.9998	0.9998	0.9998	0.9998	0.9998	0.9998	0.9998	0.9998

Index